THE CYBER EQUALIZER

THE CYBER EQUALIZER

The quest for control and dominance in cyber spectrum

By

Louis M. Giannelli

To order additional copies of this book, contact:
Xlibris Corporation
1-888-795-4274
www.Xlibris.com
Orders@Xlibris.com
116457

Table of Contents

To Cara, my friend and my muse.

Though unbeknown to you, the example of your industriousness generated the momentum leading to my commitment in writing this book.

Preface

The quest for control and dominance in cyber spectrum is a battle of brains, and in this martial arena, cyber knowledge is the ultimate equalizer. Whoever acquires cyber knowledge, and applies it so as to obtain the desired goals of a particular agenda (personal, corporate, regional, and national), obtains power and control over his nemesis, regardless of social-economical status, or the availability of a large scale technical or industrial infrastructure.

This book addresses the fact that a single individual, armed with very little capital and material resources, can achieve control and dominance over a targeted network, thus becoming a threat to such network, despite the fact that this network may have the massive technological and industrial support of a nation. In the realm of cyber spectrum a cyber David can defeat a cyber Goliath with a small amount of binary code injected inside the Goliath's brain. No amount of financial and industrial resources can protect against the power of cyber knowledge.

Those organizations who are satisfied with the pursuit of the traditional paradigm of intrusion detection *a posteriori* are doomed to spend all their time in postmortem, after the facts reactive efforts. Nobody can win a war by simply counting dead soldiers. And yet, many organizations are complacent in following this paradigm, characterized by a reactive pattern, and spending enormous amounts of resources (both human and financial) in this self-defeatist approach. Such organizations cannot conceptualize the seminal principle that cyber knowledge and cyber strategies, when designed by experienced and qualified cyber professionals, lead to preventing such casualties, while simultaneously enabling the organizing and implementing of a sensible cyber defense posture, tailored to the threat at hand. Consequently, such defense posture paves the way into a sensitive degree of success in achieving cyber superiority.

Spokespersons advocating the defeatist school of thought based on the reactive cyber defense paradigm are so used to their cyber security failures that even mock the proactive and preventive paradigm sponsored by this author as a "Hollywood fantasy". In their ignorance they cannot believe that through cyber knowledge and experience a qualified cyber defender can actually anticipate, and even prevent, an imminent cyber attack whose early indicators are detectable via forensic analysis of raw network traffic data.

Reactive cyber defense, with its complete dependence on automated systems, and limited to the fallacy of their own configuration files, will never lead to cyber superiority. The addition of cyber strategies based on knowledge of one's own cyber weaknesses, coupled with preventive strategies designed by the aforementioned qualified cyber professionals, will indeed lead an organization to afford a better opportunity of achieving a degree of cyber security and cyber superiority.

The proactive and preventing paradigm sponsored by this author is substantiated by facts and events properly documented, including cases of proactive detection and neutralization of botnets attack, undetected by the traditional automated intrusion detection systems.

In the cyber realm not every conflict between two opponents becomes a battle of brains; if one of the contenders lacks the required cyber knowledge to engage the adversary, the conflict becomes an asymmetrical carnage instead.

Introduction

The English lawyer and philosopher Francis Bacon (1561-1626) was perhaps incognizant of the exponential implications of his most famous quote "Knowledge is power". When he uttered such axiom the concept of the cyber spectrum had not yet even been envisioned. However, there is no other environment where Bacon's axiom is more pertinently applicable.

Cyber knowledge enables any human being with a truly universal power that transcends geographical and political boundaries. Cyber knowledge is the true equalizer; whoever yields such power needs no riches, or military might. When we created the cyber spectrum we literally opened the mythical Pandora's jar[1], thus replicating the disastrous event of ancient Greek mythology when Πανδώρα[2], the first woman, became the giver of all, from πᾶν meaning "all" and δῶρον meaning "gift."

The creation of cyber spectrum is irreversible, and so is the transformation that cyber spectrum has brought into human dynamics. Cyber activities are so intimately infused into the fabric of our modern society that the slightest change in the degree of cyber availability plunges our modern society into variable degrees of panic and despair. We have surrendered our most precious and intimate aspects of our personal, professional and corporate lives into the cyber realm, and any variation experienced by the performance level available within the cyber spectrum has a significant impact into our private, national and corporate existence. Retroactively we realized what an amazing and powerful "gift" we have unleashed, and we feel almost

---

[1] Commonly mistranslated as "box"
[2] The Pandora myth as narrated in lines 560-612 of Hesiod's poem in epic meter, the Theogony (ca. 8th-7th centuries BC)

powerless to control the concrete and physical manifestations of this extraordinary cyber "gift".

Knowledge and vigilance are the only two real defenses against the plethora of unforeseen manifestations of the cyber spectrum in which we now live. Only cyber knowledge and cyber vigilance stand between control and chaos. The cyber equalizer renders everyone equally vulnerable and equally powerful.

Chapter 1. The Electromagnetic spectrum

The electromagnetic spectrum (EMS) is the generic scientific term encompassing a vast array of different types of radiation. Radiation is energy that travels and disseminates in our physical environment, such as visible light from an electric lamp and radio waves from a radio transmitter, just to mention a couple of electromagnetic radiation types. Other examples of electromagnetic radiation are microwaves, infrared and ultraviolet light, X-rays and gamma-rays.

Consequently, we can state that the EMS is a continuum of all electromagnetic waves arranged according to frequency and wavelength. EMS can and does co-exist with the cyber spectrum, but they are and remain two distinct spectra. Cyber spectrum utilizes the EMS as a carrier, but even when cooperating, these two spectra remain different and distinct from one another. There are military entities that ignore this distinction and attempt to merge one into the other. EMS is not included in the cyber spectrum; the former is essentially an array of radiation waves, while the latter is essentially a collection of binary codes operating and traversing computer systems, whether as stand-alone or networked entities.

The clear distinction between EMS and cyber is unceremoniously blurred by the thousands of spokespersons representing industrial and governmental organizations. They pretend to understand the cyber spectrum, while displaying an appalling ignorance on the true nature of cyber, as evident in the uninformed statements they release to the public. Regretfully, the listeners assume these spokespersons have the technical knowledge and correct lexicon when releasing their statements, simply because they represent an institution deemed to be responsible for cyber issues. Disappointedly, their unenlightened statements become the foundation for endless repetitions and regurgitations of such uninformed statements. There are those who incorrectly define the cyber domain as the entire electromagnetic spectrum. There is, indeed, an undeniable synergy between the cyber spectrum and the EMS, since both spectra cooperate toward a single goal, that is, the transmission of information and commands. And yet, even within this synergy and cooperation, both spectra retain their unique nature and characteristics; EMS remains as the domain of waves, and cyber remains as the domain of cyber code as digital information.

The distinction between EMS and cyber exists beyond the simple realm of semantics. The distinction between the two domains is critical when it comes to encountering a cyber conflict. A leader confronting an adversary must know at all times what kind of weapons are targeting his AOR, and must know at all times what human and technical resources he must utilized to launch the adequate defenses and counteractions. If we don't properly identify the kind of weapons used against our AOR, we are doomed to waste time and resources in applying the wrong response, if any at all.

I was recently and personally involved on a case of a cyber incident lasting 22 days without resolution, and lacking the proper assessment and response. The individuals working on the case were making assumptions on the nature and cause of a suspicious network traffic, considered the result of a cyber attack. Such assumptions, however, were not based on analysis of the binary evidence. I offered to provide an analysis of the available forensic evidence that immediately revealed the network activity was not the result of a cyber attack, but rather the result of a system misconfiguration, attributed to an inexperienced individual responsible for the operation of the system in question. This is a rather typical case of assigning responsibility for cyber operations to individuals lacking the proper technical cyber knowledge required to configure the system.

The distinguishing characteristic of cyber spectrum activity is the generation, processing, and dissemination of binary code.

The dissemination of cyber code may utilize a medium other than network cables, but even when the disseminating medium may be radio waves, the cyber code remains binary code. The essence of the difference between EMS and cyber spectrum is the primary unit of information. EMS generates radio waves, while cyber spectrum generates cyber code. This distinction is quite frequently blurred by individuals attempting to define groups of activities within the generic umbrella of electronic operations, and the blurred distinction is quite often the result of using improper terminology, along with the generation of incorrect methodologies, especially in the area of cyber defense.

Modern warfare doctrines have recognized that any current and future conflict will certainly take place as a quest for dominance in the EMS, with a vast span of weapons operating in this spectrum, thus giving birth to the often confusing term "electronic warfare" (EW), a description of all offensive and defensive measures launched and counteracted by all adversarial players engaged on a declared or undeclared conflict, with each party attempting to target, exploit, disrupt, deceive, degrade, damage, and eventually destroy the electronic infrastructure of the adversaries, while retaining their own capabilities in the EMS. Since there is an international consensus in defining EW within the real of the EMS, then clearly the cyber spectrum is excluded from the EMS, in the same manner and for the same logical reason that EW excludes cyber conflict. Why? Because EMS is the realm of radiations, and the cyber spectrum is the realm of binary code generated, processed, and disseminated by computers and computerized embedded systems. Radio waves do not equate to binary code, and vice versa. This clear distinction does not preclude the cooperation of these two spectra during a conflict, but either spectrum remains unique in their corresponding essential unit of activity. When designing, planning and executing a particular defensive or offensive strategy, this distinction must be taken into consideration. Not doing so will result in negative results, and wasted time and resources.

The misunderstood equation of EMS and cyber is perhaps better illustrated by an erroneous entry in one of the most popular fora, the Wikipedia. Under the title "cyber electronic warfare" the contributor defines it as "any military action involving the use of electromagnetic energy to control the domain characterized by the use of electronics and the electromagnetic spectrum to store, modify, and/or exchange data via networked systems and associated physical infrastructures."[3]

Electromagnetic energy and EMS do not store, modify or exchange data via networked systems. Data may travel through the EMS using it as the carrier media. But when it comes to data in digital form, this becomes binary data, and as such, is no longer either "electromagnetic" or part of the EMS. Cyber, as binary code, is no longer part of the EMS. Therefore, we no longer can define cyber warfare as the use of electromagnetic energy.

> The cyber spectrum is the realm of binary code generated, processed, and disseminated by computers, whether as stand-alone systems, or interconnected in a network.

The term "cyber" is etymologically linked to the Greek κυβερνήτης (kybernētēs), meaning steersman, governor, and pilot. Now ubiquitous, the term has become a universal means to describe all activity associated with computers, information technology, and the internet. The United States government recognizes the interconnected information technology and the interconnected network of information technology infrastructures operating across this medium as part of the US National Critical Infrastructure. The fusion of these two entities occurs at the non-scientific and non-technical level, where programs are managed and funds are allocated. The EMS programs have a substantially longer history than those emerging in the realm of the cyber spectrum, and administratively it have been easier and more expeditious to merge the latter into the former, in order to appropriate the funds required for the relatively newer cyber spectrum programs.

---

[3]   http://en.wikipedia.org/wiki/Cyber_electronic_warfare

Thus, this artificial fusion remains an adaptation whose reality exists only in the minds of personnel dealing with appropriation funds. However, at the scientific and technical level, these two spectra maintain a separate existence since they are essentially different.

Chapter 2. The cyber spectrum

There are many areas where the EMS and the cyber spectrum merge, but yet without fusing into a single environment. These two may cooperate in a joint operation, but they remain distinct. This distinction is often blurred among those who are not aware of the peculiar and unique characteristics of the cyber spectrum.

The essence of the cyber spectrum is defined by the digital binary format of the data and the network protocols that transfer that data from origin to destination, regardless of whatever different supplemental and transient media such binary data may traverse through intermediate steps between the original and final destination.

The definition of cyber spectrum provided by this author is not only to retain good writing etiquette, but is also to offer the reader what is very rare to find. I have scanned, literally, not figuratively, hundreds of official publications mentioning "cyber spectrum" without ever defining the term. Good writing protocol dictates that every specific term addressed on a document should be properly defined as early as possible in the written document.

The lack of specific definition for the term cyber spectrum is quite symptomatic of the problem at hand. Our collective awareness demands the use of this term, given its critical place in our current history, and yet, it is also clear that numerous authors use this term as a buzz word, a term they cannot afford not to use, even though they do not know exactly what it means. The exact same problem affects the other buzz term thrown around in numerous publications, namely, cyber space. These two terms are derivative of cyber, and if we don't know how to define cyber, it's absolutely ludicrous to used derivatives of the term without defining the root. The bottom line is this: not everything electronic is cyber, and the electromagnetic spectrum and cyber spectrum are two different realms. Those who blur and attempt to fuse these terms do it in ignorance and in doing so they mud the waters in our search for the right answers.

The only way to find the right answers is to look in the right places. In the cyber spectrum we must know where to look during our search for the right answers. During a cyber confrontation the opponent who knows his cyber environment, and studies his adversary's cyber environment, is the one with the best opportunity for success. The research and the study of one's own cyber environment, as well as the opponent's, begins and ends with the search and study of the binary data collected during the cyber confrontation. This is where cyber forensic research becomes the foundation for cyber victory, but cyber forensic can only be conducted by human beings, not but automated systems.

One does not become a cyber warrior by simply using a computer. Just like a warrior in traditional warfare is the one who has been trained for warfare, so is the case in cyber spectrum. Trained and experienced cyber warriors do not grow on trees, and do not flourish overnight either. True and successful cyber warriors are the product of specialized study and training, individuals who do not enter into this cyber spectrum field unadvisedly. These cyber warriors must have a serious and sustained commitment to dedicate their lives to live and fight in the cyber environment, and to the study of the demanding and numerous cyber disciplines required to understand and survive a cyber conflict. Those who attempt to enter into the cyber spectrum without the commitment and the dedication described above, remain forever in the outskirts of the cyber spectrum, always pretending, never truly belonging.

Very recently I had the opportunity to see an example of this peripheral existence. I attended a cyber presentation conducted by two representatives of a government organization charged with performing an important role in countering cyber threats. The representative who took the podium proceeded to make gratuitous assessments regarding the attribution of Stuxnet. After a few minutes of listening, I raised my hand to make the following comment:

Comment: Sir, in order to observe indicators regarding potential attribution for a particular code, one must first examine the source code, if available, of the software in question. Do you have the source code for Stuxnet?
Answer: Yes, I do, and we are examining it very carefully
Comment: No, you do not have the source code. Presently, only the developers of Stuxnet have the source code. What you may have is an example of the compiled binary code of one of the variants of Stuxnet. But since you do not know the difference between source code and compiled code, your statements are questionable.

Following this short exchange the second representative, who had remained seated in the audience, stood and reply to me: *Sir, you are correct, we do not have the source code for Stuxnet, but we are doing reverse engineering on the compiled code we have obtained.*

This episode is quite typical in cyber exchanges as those taking place through numerous conferences addressing cyber spectrum activities. Organizations send representatives exhibiting very diverse technical knowledge level, and regretfully, most of the time the less knowledgeable representative acts as the spokesperson, while the one with the advance cyber technical knowledge remains in the shadows. Life in the cyber spectrum is neither an oligarchy nor a democracy; life in the cyber spectrum is a technocracy. Your hierarchical position doesn't have any meaning without the equivalent technical cyber knowledge. In cyber spectrum technical cyber knowledge rules; rank doesn't. When we speak of what we don't know, we mislead our audience.

Is this a problem only at the academic and semantic level? No, this problem affects us very deeply in our cyber security posture, and at all different levels. When we do not know where to look, we will never find what we need to know in order to present a sensitive defense before our opponent, and this adversary will always have the upper hand in the confrontation. The cyber battles we lose are not because the adversary is superior; we lose because we are so inferior in cyber knowledge and technical experience. The critical problem is that when we are incognizant of the true nature of an attack vector, we waste time and resources in finding the right answer to counteract the threat, whether perceived or real.

For example, it is quite common for individuals to seek help and advice on what they believe is a "cyber issue", but when asked about the particular details of the issue, they proceed to explain that it has to do with jamming or satellites issues. When an operation is conducted by using radio transmissions, computer networks, and satellite communications to transmit and relay information for analysis and strategic planning we are in the presence of a cooperative joint environment dedicated to disseminate intelligence. Though computer networks are involved in this mission, this is not a case of cyber operations, because the networks are simply a link in this joint environment dedicated to information dissemination. When an opponent targets the computers, as a single entity or as a whole network, by using malicious binary code as the attack vector, then and only then this scenario becomes a cyber operation. Offensive disruption of the communication links by means of electromagnetic attack vectors is a case that should be handled by electronic warfare (EW) specialists. Offensive disruption of binary code activities should be handled by qualified cyber specialists.

In the case above we can see the dangerous consequences of not being able to understand the difference between the EMS and the cyber spectrum, and the failure to recognize the difference has the potential of becoming the decisive factor between mounting a successful countermeasure or defeat. This is the reason why I previously stated that knowing the difference is more than just an academic and semantic matter.

Wireless network are perhaps one of the best examples of cooperation and joint operation between two different and yet supplemental spectra; the binary data originates in a computer, and it is prepared for network transfer by network protocols, but the intermediate vehicle of transportation between origin and final destination is provided by radio waves, acting as the carrier for the binary data. Here two spectra cooperate in a joint operation, and yet they coexist without either the cyber spectrum or the EMS spectrum altering their corresponding essence. The latter is simply the carrier of the former.

In 1895, just a few decades after the invention of the telephone, Marconi demonstrated the first radio transmission from the Isle of Wight to a tugboat 18 miles away, and radio communications was born. The early radio systems transmitted analog signals, but present day's radio systems transmit digital signals composed of binary bits. The first network based on packet radio, ALOHANET, was developed at the University of Hawaii in 1971, and enabled computers located at seven campuses spread out over four islands to communicate with a central computer on Oahu via radio transmission[4]. In this joint operation the binary data generated by the computers remains binary data (cyber data), while the radio transmission acting as the carrier remained in the realm of EMS as radio waves. This illustrates the cooperation of the two spectra (EMS and cyber), while they both retain their characteristics and identity as two different spectra. The illustration below shows how binary data retains its essence as binary data, regardless of the carrier use to transfer that binary data.

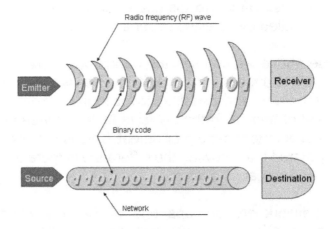

Cyber spectrum is both intangible and concrete, at the same time. The binary code that is literally created by a cyber programmer begins its existence as a collection of structured instructions to the mathematical processing unit in the micro processor of a computer, and as such they are ethereal and ephemeral, but the effect of such binary code once processed and implemented through a binary application, on a single computer or a network, becomes material and tangible in its effects.

Cyber data, though ethereal and ephemeral, does produce kinetic and tangible effects once such data has been processed. Perhaps the most infamous example of this kinetic effect is a case of malware. A set of binary instructions designed to alter or destroy other sets of binary instructions translates into a degraded or negated operation of the systems affected by the malware. When an operation sustaining a program or a plan associated with people and devices experiences degradation or a disabling condition, then the binary instructions, after having been processed and implemented, become consequences impacting the physical realm. Though cyber spectrum manifests itself in a concrete manner, it remains ethereal and intangible as binary data.[5]

There are hundreds of articles written on the subject of kinetic and non-kinetic means, available as inventory arsenal items to aid the planning and execution of operations during a conflict. Many of these authors, however, tend to use the non-kinetic term as a euphemism for electromagnetic activities that include cyber operations.

While cyber is based on binary code, and as such ethereal, the ultimate effect during the execution of this binary code exhibits in many cases kinetic characteristics. The ethereal binary code in a software program is always designed to interact with the physical components of the computerized systems where it resides. Thus,

---

[5]   Some organizations, lacking the proper understanding of the essence of cyber spectrum, speak of cyberspace as a physical domain, including satellite communications, thus artificially creating a hybrid entity by attempting to fuse the EMS and the cyber spectrum

when a user presses the "Send" key on his email program, the binary code responsible for the email operation will instruct the network card in his computer to transform the email text into digital code that will be sent through the network cable. The message will travel as a sequence of electric voltages, radio frequencies, or pulses of light. This is the kinetic effect accomplished by the email binary code written in the email software. Once these electric voltages, radio frequencies, or pulses of light arrive at the destination computer, they will have the capabilities to cause another kinetic effect, especially in the case of malicious binary code.

The most contemporary and publicized illustration of this kinetic effect in cyber is represented by Stuxnet. Its binary code was designed to modify the program of Siemens industrial control systems (ICS) by modifying the binary code on Simatic programmable logic controllers (PLCs), responsible for controlling the frequency converters, thus injecting a negative impact on the production of low enriched uranium.[6]

The goal of this chapter is to enlighten the reader into the proper understanding of the uniqueness of the cyber spectrum. While numerous uninitiated spokespersons blur the distinction by placing cyber activities in the same set of electromagnetic activities, they failed to understand that there are many cases where a cyber operation is conducted with a weaponized binary code, designed to conduct intelligence collection on the adversary, or degrade and denied his cyber capabilities. When we are in the presence of a weaponized binary code used for an offensive mission is when we are in the exclusive realm of the cyber spectrum. As stated before, the cyber spectrum may coexist with the EMS spectrum, but they never fuse, because they are incompatible in their essence.

Once again, and in order to illustrate this very common misconception, let us review a few recent articles that blur the

---

[6]   See Symantec's "W32.Stuxnet Dossier", by Falliere, Murchu, Chien, Version 1.4, February 2011, and ISIS report's "Stuxnet Malware and Natanz", by Albright, Brannan, Walrond, February 15, 2011.

distinction between EMS and cyber. A recent article on the aviation industry states that new planes featuring non-kinetic capabilities will include cyber, electronic warfare and ISR technologies.[7] Another journalist quotes a naval officer as expressing his greatest concern caused by the non-kinetic capabilities exhibited by China, leading to the wielding of dominance in the EMS, including cyber activities. The naval officer is quoted as visualizing the next armed conflict taking place predominantly in a non-kinetic EMS, with attack vectors including computer malware and jammers.[8]

These brief examples of the widely disseminated lack of cognition on what constitutes the fundamental difference between EMS and the cyber spectrum are more damaging than a simple lexicon error. Non-kinetic capabilities and effects are broader than electronic activities, and malicious binary code does not belong in the same category of electromagnetic waves, as in the case of jamming activity.

The cyber spectrum is the continuum of all cyber activities generated and executed via binary code. Operations in cyber spectrum are always very specific, because they are orchestrated by the specific goal present in the binary code. Identifying this goal is one of the most important tasks when examining the code of a cyber attack vector. Therefore, when untrained and inexperience individuals blur the demarcation line between EMS and cyber, they will always misidentify the required countermeasure.

It is highly ironic how the Western world is continually complaining about the perceived cyber prowess of China, and yet the Western

---

[7]   "U.S. Planes Will Feature Non-Kinetic Effects", David A. Fulghum, Jan 10, 2011, http://www.aviationweek.com/aw/generic/story_generic.jsp?channel=aerospacedaily&id=news/asd/2011/01/10/07.xml&headline=U.S.%20Planes%20Will%20Feature%20Non-Kinetic%20Effects

[8]   "Forget about China's missiles and stealth fighter; worry instead about 'non-kinetic' combat, Robert Haddick, January 19, 2011, http://smallwarsjournal.com/blog/forget-about-chinas-missiles-and-stealth-fighter-worry-instead-about-non-kinetic-combat

strategists, both in the military and in economy and industry, are blindly infatuated with Sun Tzu, the overrated Chinese military strategist. Only as a humorous rhetorical inquiry, I ask my readers: are we collectively infatuated with this oriental strategist because we have a complete lack of western strategists to emulate, in thought and in actions? Don't we have anyone in Western civilization history since the sixth century BC to learn from, and consequently, in total destitution, we have to look at the oriental adversary we seem to fear the most? If you as the reader reply in the affirmative, we are all in a very sad state of affairs . . .

In a cyber conflict we deal with an adversary who has written or modified a binary code with a specific offensive goal. To know the code of the attack vector is the foundation for any cyber countermeasure, and possibly the beginning of the path to victory, or at least a successful cyber defense strategy. For those who follow Sun Tzu, I'd like to remind them to also follow one of his most quoted dicta: "know your enemy and know yourself."

I don't pretend to be a strategist, but I do know something about studying your cyber adversary, after encountering him on a daily basis, during a decade of active and proactive cyber defense protecting government networks. Learning about the strategy and the cyber tools used by an adversary leads to the learning of your adversary's offensive plans, and using trend analysis and anticipatory analysis it is possible to present a successful defense, and survive to fight other adversaries the next day.

Chapter 3. The Age of Innocence

Once upon a time there was a small number of computers that were independently providing services to small scientific communities in the USA. But the inherent nature of scientific exchange required for the advance of science via open communication of discoveries led this small community to envision a system of establishing a line of remote communication among this small number of computers, and the Internet was born. The name "Internet" represents the contraction of the terms "interconnected networks".

The set of communicating rules facilitating this remote connection between computers are known as "protocols", a series of binary code programs designed to facilitate the numerous ways of establishing remote connections and the exchange of data between the interconnected computers. However, this set of rules (protocols) were designed and implemented with only one goal in mind; to facilitate the communication among the scientific communities exchanging data, and since there was an intrinsic trust relationship among the participating scientists, there was no need to integrate security measures in the design of the binary code supporting these networking protocols. This was the age of innocence. These were the days of living in the Land of Oz minus the Witch. The small scientific community was happily prancing through the fields of Oz, enjoying their network exchange in perfect and idyllic form. There were no dangers of any kind.

Paraphrasing Dickens[9] we could say that for the nascent Internet community "It was the best of times, it was the worst of times, it was the age of wisdom, it was the age of foolishness." This dichotomy was created because the protocols written for the implementation and deployment of the Internet would allow it to

---

[9]   The famous 19th century novel "A Tale of Two Cities".

flourish and bloom into the global system we all enjoy today, but these very same protocols contained the seeds of the defeat and destruction characterizing the current pandemic status of chaos and disruption, caused by the inherent vulnerabilities built into these protocols designed with a total lack of protection. To illustrate this point let us mention the most vulnerable protocol and at the same time the most popular and virtually indispensable SMTP protocol. This particular one is the work horse of our pervasive dependence on email communication. The SMTP protocol lacks the most rudimentary security features, thus facilitating the most pernicious forms of cyber attacks such as spam, spoofed, and spear phishing emails. Thus, the Internet brought the best of times, while at the same time planting the seed for the worst of times, and while it ushered the age of wisdom exchange, it also subjected us to the foolishness of carrying within itself the seed of destruction.

This early situation has not changed. The scientific community remains unaware of the many and vicious dangers lurking in the Internet at large, and the World Wide Web (WWW) in particular.[10] The same is true with the rest of the population at large, and all these unaware users contribute to the explosion and dissemination of malware. Most of the times, when living in this state of cyber innocence users are dramatically surprised when they fall victims to a cyber attack. A dear friend of mine, after becoming a victim of a drive-by attack, and during the time I was working on sanitizing her computer from the infecting malware, expressed this feeling of despair by saying: "I feel so betrayed". This is the exact representation of living in the age of cyber innocence,

----

[10] The Internet, the global infrastructure of interconnected networks, should not be confused with the WWW. The latter is only a subset of the former, being just a large collection of web servers kept operational within the larger infrastructure provided by the Internet. The majority of the WWW users ignore this important distinction, thus giving origin to the erroneous question "Is the Internet down?" when they cannot connect to a particular web site. The Internet is never down, but a particular web site may be down at any given time.

and experiencing the rudely awakening, and the accompanying trauma, of becoming a victim of cyber predators. We become users of the WWW, and we are expecting a relationship of trust among the community of users. We navigate through the WWW on the basis on an implicit trust relationship, and when we fall victims to a cyber attack, it's very difficult to recover from the shock that we cannot truly trust anything that is offered in the WWW.

This author does not advocate to transition from a state of cyber innocence into a state of cyber paranoia. Rather, this author advocates for adopting a healthy and cautious lifestyle when navigating the cyber arena. Perhaps we should re-learn the important lesson we traditionally teach our children when we tell them not to open the door to strangers, and refrain from accepting candy from them. And yet, when traversing the streets and neighborhoods in the Land of Cyber-Oz, we forget the terrible presence of the cyber Witch, and we proceed to open our cyber doors to all type of strangers, while accepting cyber candies from them as well. When traversing the cyber arena we have to adopt a healthy measure of caution and alertness, on an increasing escalating progression as we move from familiar to unfamiliar cyber territories.

It is this implicit trust that is being exploited by the malicious actors lurking in the cyber highways of the WWW. They are counting on the trusting nature of WWW users in order to create the scenarios required for the dissemination of their cyber exploits. A constant state of alertness is what is required on the part of WWW users in order to navigate to their desired web site, while staying outside the reach of cyber predators. This state of alertness may represent the difference between becoming the victim of a total infection, as opposed to a partial infection. My friend possessed that state of alertness, and the infection of her computer, while serious, did not reach the fatal state of total destruction. As soon as she realized something out of the ordinary was taking place on her computer, she terminated the connection to the malicious site, and even though her system was partially disabled, I was able to sanitize it and restore it to a functional level.

Is there a perfect "check list" that may allow a user to know when something is going wrong? No, there isn't. The only safe procedure to follow is to remain in a constant state of alertness while navigating the WWW. How do we accomplish this state of alertness? By drawing some basic principles for cyber safety, and avoiding erroneous assumptions. Let's just mention a few of them.

Erroneous assumption 1. *I don't have anything to worry about my computer. It's well protected because I have anti-virus (AV) software installed in it.*
Cyber safety principle 1. AV software is not a panacea against all cyber exploits. There is no single AV capable of detecting all malicious binary code designed as a virus. Furthermore, not all malicious code comes in the form of a computer virus. Not every AV maintains a current set of signatures capable of detecting virus. AV software will not protect, among others, against phishing, spear phishing, and drive-by malware attack. AV software will not protect against a denial of service (DoS) attack, or a botnet-driven distributed DoS (DDoS) attack. Even analysts in the IT security market err in making this erroneous assumption, and they publish their assessments by condemning the fact that AV products are not catching all the online exploits. This is an unrealistic expectation, the product of ignorance regarding cyber security principles.

Erroneous assumption 2. *I don't worry about zero-day exploits because I have anti-virus (AV) software installed in my computer.*
Cyber safety principle 2. AV software is not designed to offer protection against zero-day exploits, because there is no correlation between a zero-day exploit and a computer virus signature. A zero-day exploit is a vulnerability existing on a legitimate application, but not yet detected by the code developer, or not yet "patched" by the application vendor. A computer virus can be detected and nullified by an AV signature only after the said virus is detected, identified, and counteracted.

Software researchers dedicated to analyze binary code may discover dormant vulnerabilities. The discovery may be shared with the vendor responsible for selling the vulnerable application, or the researcher may decide to write a piece of code to exploit the newly found vulnerability. The final decision hinges on the working ethics of the researcher. All software applications carry undiscovered vulnerabilities, simply because of the nature of business and marketing. Any software developer has a deadline to complete the binary code for a given application. The deadline is driven by business and marketing requirements. During the finite time available for developing the binary code there is also a finite window to test the new application. Business and marketing demands and deadlines must be met in order to obtain a reasonable profit from the new application.

On the other hand, a virus is not an undetected flaw in the binary code of a new application. The term computer virus was derived from the field of biology, where a virus is defined as a parasitical organism that attaches itself to a host organism in order to replicate itself, and initiate the infectious and destructive dissemination within the host. Similarly, a binary virus is a parasitical computer program that requires a cyber host with a suitable application to attach itself, and begin the destructive sequence of replication and dissemination. Thus, the difference between a zero-day exploit and a virus infection becomes obvious. A zero-day exploit is a dormant flaw in the original binary code of a legitimate application, while a computer virus is a malicious binary code designed to be injected into a legitimate application. This is why an AV cannot provide protection against a zero-day exploit.

Regretfully, this clear distinction is not understood by many of the numerous unqualified individuals introduced into the cyber arena via cyber metamorphosis.[11] I have seen too many reports where these individuals report and lament a zero-day infection that took place "despite having AV protection installed on the infected systems."

The number of cyber exploits is continually growing and expanding, and there is no single cyber security product capable of defending against all of them. The Maginot mentality did not prevent the

_____
[11]   See chapter 4 ahead

German invasion of France, and AV software will not protect users against all cyber exploits. What do we need then? Users need to abandon their utopian view of the WWW as a Witch-less Land of Oz, and wake up to the reality that the WWW is a dangerous jungle, and any survival is dependent on how cautious and shrewd we are in our walking around this dangerous environment.

When visiting familiar web sites, the user should study their cyber behavior, their appearance, and their "feel". When detecting a level of abnormality on these elements, proceed with caution or disconnect from the site. When visiting unfamiliar sites, cautiously assess their behavior in order to learn their normal functionality, but remain alert and ready to leave if you perceive a threat. By disconnecting in the case of a perceived threat there is nothing to lose, since you can always return. But if your caution saves you from becoming victim of a cyber exploit, you have much to gain.

When confronted with a cyber attack, every user has a defensive action in his power, regardless of his level of technical cyber expertise. As soon as the user perceives something suspicious and potentially dangerous, such user should disconnect immediately from the site originating the perceived attack. After this immediate defensive reaction there will always be time to seek the assistance of a qualified cyber professional, assess the damage, and restore the threatened system to its normal operational status. Be always ready to detect a potential threat, and disconnect. This is not the time to freeze and panic. This is the time to react against the threat by terminating the connection and shutting down your system. No amount of malware can take complete control of your system when the power is shutdown.

Once upon a time there were a small number of us connecting to the Internet, not the WWW, because these were the times when there were no web browsers. I know, I know, many of you must be asking yourselves: "what is he talking about"? How did we manage without a web browser? Those were the days when we literally converse with our computers by actually typing commands on our keyboards. Buy why, you may ask? Didn't you have a mouse to click some icons on the desktop and tell your computer what to do? No, we didn't have a mouse, and through commands typed on the keyboards we instructed the computer on what to do and where to connect, and we where able to connect with other computers and servers, and download documents and pictures, but there was a lot of typing in order to specify exactly the document to download, and type the location where such document or picture was located. And the number of those connected through the Internet was only a minute fraction of the number of users connected today through the WWW, the playground of millions and millions of users navigating this graphical world where everything is just one click away from us.

And then, all of a sudden, the small sand box in which we used to play became a huge field, a virtual global village, where everyone from around the globe came to play, and to prey on each other. Perhaps the virtual proximity brought by the WWW is what causes us to live in this illusionary world we still perceived as a small sand box, and we assume we are still in our backyard, playing in our safe and protected sand box, and that's why we jump into this sand box without a care in the world. Everything is within our reach, and that virtual proximity induces in us this apperception of innocent behavior while we navigate the WWW.

We collectively and individually exhibit the same degree of blind innocence when we implicitly trust that the financial and personal information stored on large databases is adequately protected. That degree of trust is undeserved, since all databases are exploitable, without exception. Is this a pessimistic or a cynical statement? Neither, since this is a technically and empirically realistic assessment. Databases are capable of providing a degree of integrity and confidentiality that is proportional to the degree of security implemented in the configuration assigned to the authentication procedure implemented to interact with the database.

I have personally examined numerous cases of databases compromised by online predators who took advantage of the gaping security holes left in place by irresponsible and negligent database administrators. There are plenty of security measures available to protect databases, but ignored by their administrators. I have personally assisted database administrators to successfully implement these available security procedures after their databases were compromised, with excellent results, thus preventing future exploits.

It is highly irresponsible to allow an unprotected database to go online, because we have the means to protect them, but the pandemic of compromised databases continue to escalate. Is this the result of extremely gifted cyber predators? No, a thousands times no! The rise on the escalation on compromised databases is caused by the crass ignorance and negligence of careless database administrators that decline to apply the known and available protective measures during the configuration of the database. Cyber predators continue their successful exploitation simply because database administrators continue neglecting the implementation of the defensive means readily available. The exploitation of an unprotected database is a trivial matter that requires very little effort. Likewise, the protection of a database requires very little effort as well.

SQLIA[12] has remained among the 10 top cyber exploits for the last five years simply because of the great success rate in exploiting online databases. Humans interact with a database through the Structure Query Language (SQL), a language of queries designed to manipulate the data stored in the database. When a database administrator fails to configure his database with precise instructions on how to discriminate between authorized and unauthorized user inputs, that database is doomed from the very moment it becomes available online. A cyber predator can use unauthorized SQL queries at will to exploit the contents of the database, and this type of exploit has become so trivial that there are automated SQLIA tools available everywhere online, so that

---

[12]   SQLIA is the acronym coined by this author on professional papers written to increase understanding and awareness regarding this pervasive exploit. It stands for Structure Query Language Injection Attack. The term was coined on "Detection and Mitigation of the SQL Injection Threat." NTC Paper, November 2006

even novices can mount this type of attack. Next time you read or hear in the news that credit cards numbers and PII[13] have been stolen, don't assume that the cyber criminals are so smart that they defeated all the database defenses. It takes very little talent, if any, to attack a database by using SQLIA when a defenseless database stands online with a target painted on it because the database administrator failed to configure the well-known and available configuration defenses.

We need to mature and become comfortable but cautious while traversing the online territory. We are no longer in our small protected sand box. Now we have entered this Gigapolis[14], this enormous city made of a gargantuan amount of data, produced and stored in its ever expanding topography. We are no longer in a safe and protected environment, but in a massive and dangerous territory. We can instantly gain access to any data in this vast Gigapolis, but we have to exercise prudence and caution on what kind of data and where we get this data. We can no longer implicitly trust everybody in Gigapolis, because we no longer know everybody who traverses the online highways in and out of Gigapolis. In our little sand box in our backyard we used to behave with implicit trust. Now, while traversing Gigapolis, we must walk with a healthy degree of caution, and we can only ascribe our trust to those who have given us a solid reason to trust them. If we are to survive in Gigapolis, we must become street savvy.

---

[13]   Personally Identifiable Information
[14]   Gigapolis is a term coined by this author, combining two Greek terms, namely, γίγας (giant), and πόλις (city), as a metaphor to represent the enormous size of the data stored online, and made accessible via the WWW.

Chapter 4. Cyber Metamorphosis

After the advent of the cyber lexicon into the collective awareness and public knowledge, commercial, governmental and military institutions initiated a migration of personnel into a revised organizational chart, metamorphosing informational technology (IT) and communications (COM) personnel positions into cyber positions. This action brought the mushroom-like phenomenon of the appearance of cyber titles attached to job descriptions sprouting over night, with a deluge of cyber entities surfacing like mushrooms after the rain. However, this renaming changed neither the composition of the organizations adopting the change, nor the professional qualifications of the personnel wearing the new position titles. The renamed IT and COM individuals were neither endowed with a new skill set to prepare them for cyber operations, nor with a new mindset to function as cyber professionals.

While the renaming in itself did not affect either the industrial, commercial or governmental cultures, a serious professional identity crisis was born. Changing from an IT or COM badge to a cyber badge does affect the professional requirements, because cyber is different than IT and COM, since the most remarkable difference resides in the skill set and the mindset required by cyber operations.

Cyber is a technocracy, not a democracy. The paramount criterion is technical knowledge and technical experience. Unqualified opinions from individuals non-compliant with the said criterion do not count. This is an important distinction required to properly articulate and understand the difference in the skill set and mindset required in the cyber spectrum. The essence of cyber is binary code, and it is within the realm of this computer language code where all cyber operations occur, whether authorized or unauthorized. The enormous amount of traffic traversing the

Internet may have a legitimate or illegitimate purpose, and the distinction between one or the other can be discerned only by those who understand this technical language. Those who attempt to enter the cyber spectrum, while lacking the proper understanding of the technical cyber language, adopt the untenable supposition of assuming that everybody can operate in the cyber spectrum.

Interacting with other users in the cyber spectrum is not synonymous with operating in the cyber spectrum itself, in the same way that ingesting a medicinal prescription does not make the user to become a physician. In the realm of the cyber spectrum there are only two kinds of entities; users and operators, where the former do not understand the essentials of the cyber language, and the latter do. Only those who are capable of developing, manipulating or understanding cyber code, or networked and networking protocols, only these belong to the operator category; all others are simply users.

The failed attempt to democratize cyber is driven by the desire to transition and metamorphize from an IT or COM environment, attempting to claim legitimacy by proxy into the cyber spectrum. Some organizations take this desire so far as to claim that everyone in front of a keyboard (virtual or physical) becomes a "cyber warrior". This is an illegitimate claim; one doesn't become a cyber warrior without becoming molded by the skill set and mindset required by cyber. One does not become an aviator by simply seating on a plane.

The cyber spectrum is very unique and cannot be adopted or morphed into pre-existing skill sets and mindsets. Those who pretend to adapt or adopt cyber into their organizational universe do violence to cyber, by attempting to force cyber into pre-existing models. Cyber has a unique model that is incongruent with other models. When this unqualified adoption and adaptation of cyber does occur, the result is technical and professional mediocrity. Let's use an example from cyber history, During a period of ten months between 1986 and 1987, a young astronomer became a true cyber warrior when he detected, tracked, profiled and identified a cyber spy intruding and exploiting educational, governmental and military networks, systems and data.[15] All the national agencies he contacted exhibited neither understanding nor interest in safeguarding the targets selected by the foreign cyber spy, the German Markus Hess, who was working in collusion with the Soviet KGB, and with the Chaos Computer Club. How much more "cyber agility"[16] do we find among the same organizations today, in 2011, 25 years after the events narrated and documented by Cliff in his book?[17]

He became a cyber warrior because he embraced the skill set and the mindset required to contend with a cyber adversary, and he designed and implemented ad hoc cyber COAs[18] to impede and lead the contender into a cyber trap. He did all this by himself, without the help and support of any of the official organizations he contacted. And he did so because he cared for safeguarding the systems and the data targeted by the adversary. Eventually he delivered into the hands of the very same organizations the cyber spy that had ravished the networks and the data belonging to educational, governmental, and military institutions.

The institutional apathy that this lonely cyber crusader encountered during the course of the cyber attacks derived primarily from three factors: failure to understand the technical scope of the cyber intrusions, failure to grasp the strategic and tactical gain acquired by the cyber spy, and the bureaucratic attitude represented by the "this is not my bailiwick" irresponsible behavior. 25 years later, have we really overcome these three negative factors? Collectively, we

---

[15]   Cliff Stoll, The Cuckoo's Egg, Pocket Books, 1989
[16]   Cyber agility is the measure of cognitive understanding of the criticality of a cyber event when a person in leadership is briefed about such cyber event by a technically qualified cyber operator.
[17]   Ibid
[18]   Course of actions

certainly hope that we have, but personally, this author continually encounters the same unchanged apathetic attitude.

Someone may argue that today we are incorporating an intensified cyber technical curriculum in the formation of cyber personnel, but this technical curriculum is simply offering an expanded set of canned tools, which are not necessarily conducive to the creation of the new mindset required in cyber. We are reaching an intermediate level of technical cognition with the said tools, but such tools by themselves will not allow us to confront and defeat the cyber adversary. Cliff had the tools, but he also had the unique mindset required to confront the adversary and fold his malicious goals.

Knowing that there is a tool available to analyze the network traffic does not automatically allow us to foresee and defeat the COA of the cyber opponent invading our network. Countless metamorphosed "cyber" assets lack the knowledge and skills to understand and interpret the amount of raw data information displayed by a network analyzer. This tool offers a wealth of valuable information in monitoring a network and allows us to see the prelude of a cyber attack, but this advanced notice of an impending cyber attack can only be interpreted by qualified and trained cyber operators, with the knowledge of networked and networking protocols, and the ability to read and understand ASCII and hexadecimal language. Can you understand and interpret the network traffic below, captured by a network analyzer tool? If you do, you are among those few that qualify as a cyber warrior.

```
18 25.290361          66.35.45.157          192.168.1.102         TCP    http > ardus-mtrns [SYN,
19 25.290416          192.168.1.102         66.35.45.157          TCP    ardus-mtrns > http [ACK]
20 25.291299          192.168.1.102         66.35.45.157          HTTP   GET /links.html HTTP/1.1
21 25.343803          66.35.45.157          192.168.1.102         TCP    http > ardus-mtrns [ACK]
22 25.345562          66.35.45.157          192.168.1.102         HTTP   HTTP/1.1 301 Moved Permar
```

⊞ Frame 20 (833 bytes on wire, 833 bytes captured)
⊞ Ethernet II, Src: Intel_76:e5:e0 (00:07:e9:76:e5:e0), Dst: 58:6d:8f:73:23:9c (58:6d:8f:73:23:9c)
⊞ Internet Protocol, Src: 192.168.1.102 (192.168.1.102), Dst: 66.35.45.157 (66.35.45.157)
⊟ Transmission Control Protocol, Src Port: ardus-mtrns (1117), Dst Port: http (80), Seq: 1, Ack: 1, Len: 779

   Source port: ardus-mtrns (1117)
   Destination port: http (80)
   [Stream index: 13]
   Sequence number: 1   (relative sequence number)
   [Next sequence number: 780   (relative sequence number)]
   Acknowledgement number: 1   (relative ack number)
   Header length: 20 bytes
  ⊞ Flags: 0x18 (PSH, ACK)
   Window size: 65535

```
0210  0d 0a 41 63 63 65 70 74  2d 45 6e 63 6f 64 69 6e   ..Accept -Encodin
0220  67 3a 20 67 7a 69 70 2c  20 64 65 66 6c 61 74 65   g: gzip,  deflate
0230  0d 0a 48 6f 73 74 3a 20  69 73 63 2e 73 61 6e 73   ..Host:  isc.sans
0240  2e 6f 72 67 0d 0a 43 6f  6e 6e 65 63 74 69 6f 6e   .org..Co nnection
0250  3a 20 4b 65 65 70 2d 41  6c 69 76 65 0d 0a 43 6f   : Keep-A live..Co
0260  6f 6b 69 65 3a 20 5f 5f  75 74 6d 61 3d 32 31 31   okie: __ utma=211
```

Only the alignment of the cyber skill set and the cyber mindset in one and the same person will allow such individual to successfully confront a cyber adversary, understand his plan of attack, anticipate his next move, and present the adequate and tailored cyber defenses required by the specific attack vector employed by the opponent. This is how a certified cyber warrior steps onto δρόμο νίκη (dromo niki). This dromonikian[19] achievement, stepping onto the victory road during a confrontation with a cyber opponent, is not a moment in time, but a process, when all the cyber skills and the cyber mindset are heightened, and shrewdness, anticipation, perseverance and sophrosyne[20] guide the cyber warrior to withstand or overcome the cyber opponent, depending on the current rules of engagement (ROE).

In cyber only the qualified operators can accurately determine when a cyber event is significant or inconsequential. However, because we have many institutions that have metamorphosed their former IT or COM personnel into the cyber category, the only factor used to determine the importance of a cyber event is quantification, and deriving from that, the threshold paradigm. This concept is based on the fallacy that only a certain amount of network connections and a certain amount of data transfer is an indicator of an intrusion. How many holes it takes for a boat to sink? Only a qualified cyber operator, with knowledge of the technical cyber language and the experience of analyzing network traffic can determine when a cyber event is significant, and quantification is neither the only nor the most important factor in making such determination. This author has detected, counteracted, and nullified several DDoS[21] attacks during the prelude actions of the nascent attack. During the prelude to a DDoS very small amount of network data is present, and dependence on the threshold paradigm will never detect the DDoS attack in progress. The signs accompanying the prelude to a DDoS attack are so small

---

[19]   Dromonikian is a term coined by this author, and it's derived from the Greek phrase "δρόμο νίκη", the victory road.
[20]   See chapter 15 ahead.
[21]   Distributed Denial of Service, a cyber attack designed to paralyze networked systems.

that they will never comply with the quantification requirements of a threshold model. If this author would have adhered to the fallacy of the "threshold paradigm" the DDoS attacks would have been successful, and they would have been discovered only after the facts, after the adversary had already achieved success in his attack.

The threshold model in itself is a positive and effective factor only when used in conjunction with, and subordinate to, a qualified cyber defender, possessing the knowledge and the experience of operating with the required cyber skill set and mindset. The threshold model, when used in exclusivity, is only a partial solution that provides a limited degree of cyber defense against known cyber threats. Many organizations adopt this model because they lack the understanding of the full scope of a cyber conflict, and because they don't have qualified cyber defenders as members of their personnel. A former IT or COM asset that has been relabeled as a cyber asset does not have either the required skill set or the mindset required to operate in a contested cyber AOR.

Cyber metamorphosis is only acceptable as a transition stage to the formation of a contingent of fully qualified cyber professionals. However, if the organization adopting cyber metamorphosis considers it as the final solution, the shallowness and mediocrity of the cyber defense plan will simply become and band aid, not a cure. Cyber defense is not a trivial task, and we should not approach it in a trivial manner. If within the resources of our own organization we cannot afford to grow and mature certified and qualified cyber operators as described in this book, then we need to look outside our organization. Cyber confrontations are not for novices, and automated cyber defenses are useless without the involvement and leadership of true cyber warriors.

Chapter 5. Cyber Conflict

The world at large is in a precarious condition when it comes to defining a state of cyber conflict. Semantics are a contributing factor since the global consciousness on this topic is greatly influenced by the faulty terminology coined by the media and their insatiable taste for speculation and hyperbolic jargon. Perhaps the greatest obstacle in defining the nature of a cyber conflict is the sensationalistic term "cyber warfare", created and abused by the media. This sad state of affairs is aggravated by the influence of the media into the policy-making and the judiciary bodies. None among this trio of vocal entities have any qualified opinion on defining what constitutes a cyber conflict, due to two main reasons. First, none among them have a technical knowledge of the essence of cyber, and second, all three of them insist on a futile attempt to define cyber from a historical perspective, failing to see that cyber does not have a historical antecedent.

There is nothing in human history that can be used as a comparative entity, because cyber is completely new, an entity on its own category, without historical precedent. Any attempt to compare cyber with traditional warfare is an exercise in futility. Since the media, the policy-making and the judiciary bodies do not consistently and methodically engage the technical advice of cyber professionals, the pontification of these three bodies regarding cyber warfare is both obtuse and misleading. The so-called cyber experts sporadically consulted by these three bodies are usually spoke persons from technical companies operating within the cyber spectrum, but they rarely have a personal and empirical knowledge of the cyber technical dimension, because they maintain a rather peripheral relation to the cyber arena. An illustration concerning this technical ignorance about cyber is the abused artificial dichotomy between kinetics and non-kinetic effects, where the erroneous presupposition is that the former is synonymous with traditional warfare, while the latter is associated to some degree with their views of cyber warfare.

This artificial distinction arises from a crass ignorance regarding the essential nature of cyber. Cyber operations may appear to the uninitiated as invisible to the naked eye, but cyber operations always have a kinetic effect, they are always amazingly fast, and they always introduce kinetic results, though sometimes unwitnessed, simply because the cyber operation may have an initial and primary impact perceived only at the binary code level. As such, the unwitnessed impact on the binary code may have delayed effects that, when eventually manifested, may not be associated with the initial impact of the affected binary code. Eventually this primary impact may migrate into an observable kinetic impact that becomes a witnessed event. The fallacy represented by the type of popular apperception described in the above illustration is one of the main hurdles in achieving an accurate legislative foundation for an actionable and relevant body of laws regulating the ROE when confronting a cyber conflict case.

When policy-makers and legislators are left unattended in matters pertaining to the cyber arena, they inevitably follow the traditional paradigm of blaming the software vendors for any discovered vulnerabilities. The cyber code written for an intended use does not remain a static entity. Once installed on a particular cyber system, that code begins interacting with all peripheral systems and the supporting network infrastructure where the code is installed. Many unforeseen interactions may occur after the cyber code is installed, and it is the responsibility of the end user on a home system, or the system administrator on an enterprise, to observe and monitor the behavior of the newly installed software. The end user is partly responsible for any vulnerability discovered once the software is installed.

An article published in 2005 proposed a new tort of negligent enablement seeking to hold software vendors accountable for defective code that allow cyber criminals to exploit "known vulnerabilities".[22] While the same article does acknowledge that users do receive "a steady stream of releases, patches, updates", the authors do sponsor the utopian view that the vendors should offer "a comprehensive software security solution prior to release".[23] These ludicrous statements reveal the crass ignorance exhibited by these two authors in cyber matters. They assume that the vendors sell software susceptible to known vulnerabilities. To prove this would require that every vendor submits the new code to a technical review board that, after performing exhaustive reverse-engineering analysis on the new code, they can certify the new code is not vulnerable to known exploits. However, as explained above, new unforeseen vulnerabilities arise only after the new code is installed and begins interacting with its particular cyber and networking environment. In the cyber sphere there is no such thing as a comprehensive software security solution prior to release. This is the area of responsibility of the end user.

While the authors of this particular tort endorse the allocation of responsibility "to both software manufacturers and end users"[24], the failure of their proposed tort is ignoring the unique nature of cyber, and the lack of understanding on how new cyber code operates and interacts within the environment where it is installed. The authors of this proposal failed to seek the guidance of qualified cyber subject matter experts (SMEs) prior to the release of their legislative proposal.

SMEs should become the de facto advisers for policy-makers and legislators in defining what constitute a cyber conflict, and in defining and assessing the magnitude of the impact of an offensive cyber operation, as determined by an analysis of the technical impact of such adversarial operation. The impact of any such inimical cyber operation is never the result of a casual or unintended action, since all cyber commands issued during an offensive cyber operation are premeditated and calculated.

---

[22] The Tort of Negligent Enablement of Cybercrime, Rustad & Koenig, 2005

[23] Ibid, p. 1557

[24] Ibid, p. 1561

Measuring this impact will provide a reliable and qualitative basis to plan the appropriate and commensurable countermeasure. The media should be relegated to a purely informative role, when and if necessary, and divested of any credence when and if the media resorts to their typical speculative and unprofessional role of attempting to misuse their public forum as a sensationalistic and self-serving tool.

Journalists, by the very nature of their professional expertise, should not be ascribed as cyber SMEs. They are not, and they can never become, an authoritative voice in matters germane to the cyber spectrum in general, and to a cyber conflict in particular. Media should be limited to their intended role as town criers; nothing more, nothing less. Likewise, the bureaucracy embedded in the policy-making and legislative machinery should not become the basis for codifying the laws defining and regulating a cyber conflict. Rather, the leading role in these matters should be assigned to certified and accredited cyber SMEs, to assist and guide the policy-makers and legislators in cyber conflict resolutions.

This author favors the term "cyber conflict" over the sensationalistic term "cyber warfare" sponsored by the media and all other organizations blindly following their lead, while uncritically accepting and regurgitating their unqualified statements regarding cyber matters. The term "warfare" is loaded with historical and preconceived notions that force a cyber scenario into a traditional definition of human hostility, and associated with legal concepts universally accepted.

As already stated, cyber is a new entity in history, lacking any antecedents, and as such it requires a new mindset to deal with the technical and legal aspects involved in a cyber conflict. This can be better illustrated by referencing the emerging of quantum computing (QC) into the cyber scene. There is no antecedent for QC in the traditional cyber arena, where everything is rooted into the binary language and binary math. In QC the traditional mindset built within the binary framework has no place, simply because QC does not evolve from the binary environment. QC is a completely new framework of computing, with a completely different set of computing principles, where binary language and binary math find themselves in foreign territory, even more accurately, in a foreign universe. To operate in the QC universe we need to learn a new set of principles by adopting a new mindset. The same is true in cyber conflict, since it requires working with a new set of principles previously unknown, and any attempt to take the easy way out by seeking to force cyber into our traditional views on warfare is a misguided and foolish effort.

Only after accepting and learning the newness of the cyber principles we may attempt to codify and legislate definitions applicable to a cyber conflict and its unique set of ROE. I have met numerous leaders that are persistently looking for a translation of cyber principles into traditional warfare. There is no Rosetta Stone, and there will never be one, available for this task. He who wants to learn cyber has to learn to think in a way that is in accordance with cyber principles, and to use a language that is unique to the cyber spectrum. Those who insist on finding a non-existent antecedent in traditional warfare, and force it into the cyber realm, are engaged in a futile and unproductive effort. Those who tried to create policies to codify and legislate cyber conflicts must learn to think in accordance with the cyber mindset, and embrace the cyber language. Professional cyber SMEs are available to provide the require guidance for this transition.

If those who are attempting to legislate do not make a serious effort to learn the essential concepts of cyber, and adopt the cyber mindset, they will continue to remain paralyzed by the uncertainty of what a cyber weapon can do. Legislators in general and legal professionals in the judicial branch of the Armed Forces service live with the constant dilemma of considering and evaluating the potential implications associated with the employment and utilization of weaponized cyber assets. This uncertainty, hovering over their consciousness like the legendary Sword of Damocles[25],

---

[25] This legend, introduced by Cicero into European literature, narrates the episode of Damocles whishing to taste the power and authority of Dionysius II, a fourth century BC tyrant of Siracusa, in

illustrates the enormous tension and responsibility of legislating over a legal topic that has potentially devastating ramifications and consequences.

A reputable aviation industry magazine recently published an interview where a member of the Armed Forces judicial branch expressed this concern, by comparing the certainty factor in using a conventional bomb, and the many unknown factors associated with the delivery of a so-called "e-bomb". Incidentally, this is a perfect illustration of the inaccurate lexicon used by leaders lacking the proper glossary when referencing to cyber topics. As a corollary, this also illustrate the resistance to learn and utilize the proper cyber terminology, since there are numerous certified cyber professionals ready to assist legislators in understanding the basic cyber terminology. This is more than just semantics; this is a problem of increasing the fear factor by attempting to address unknown issues.

And yet, this argument, though apparently valid, is inaccurate. While it is correct to assess that in the case of a traditional bomb all the damage factors are exactly known prior to the release of the weapon, this assumes that the delivery of the bomb is precisely on target, but we know that this is not always the case. The uncertainty factor is there, but we feel we can make a precise assessment of the decision to release the bomb because all the involved parties in the decision-making group can understand the quantifying parameters of delivering such bomb.

Why then do legislators want to demand that a precise set of quantifying damage assessment parameters are known and guaranteed before the authorization to release a cyber weapon? Because such legislators and policy-makers do not understand the cyber language by which we can quantify the impact of such cyber weapon, and lacking understanding, they cannot make a decision

---

Sicilia, Italia. The tyrant invited Damocles to sit on his throne, but suspended a large sword above the throne, held by a single horse's tail hair. Damocles very quickly declined the privilege of sitting on the throne.

for fear of what they think are "unknown" consequences. They want the same assurance than with traditional weapons, while the fear of what they label "unknown" factors (actually not understood by them) hovers over their heads like the Sword of Damocles.

Let's ask ourselves a fair question from our own national history. Did we truly know and comprehend all the quantifying parameters of the effects of the atomic bomb before we delivered Little Boy on Hiroshima and Fat Man on Nagasaki? The plain answer is: No, we didn't truly know. And yet, when the moment arrived, the difficult decision was made to use a weaponized form of the new nuclear technology. Did we entangle ourselves is a never-ending and paralyzing discussion on whether or not to use the new technology? No, because when no other options where available, we decisively acted, and the rest is history.

Today we are standing in exactly the same spot. Back on 1945 there were no antecedents for the use of nuclear technology as a weaponized tool. Today, we have a new technology, without antecedents on all the nuances associated with its use under the umbrella of weaponized iterations of this new technology, and while we parade in the cyber arena with a target painted on our back, we engage in endless discussions and ruminations on whether or not to use these new weapons to defend our national interests. Qualified and certified cyber SMEs are always available to explain the assessment parameters sought by legislators and policy-makers. However, the aviation magazine article referenced before states that the consulted cyber weapons guys do not have an answer to the issue of damage assessment associated with such cyber weapons. Who are these so-called "cyber weapons guys"? Are they qualified, certified, empirically knowledgeable cyber professionals? What credentials do they have? Qualified cyber SMEs do have the knowledge and means to offer precise quantifications for the use of cyber tools under a given set of ROE, but it's unrealistic to expect the delivery of this quantifications using traditional terminology, because cyber is a new science and technology without precedents. Unless legislators and policy-makers learn and truly understand the language of this new technology, they will be unable to digest the required quantifications, and they will remain fearful of what they can't understand, and unable to make decisions and write policies for the use of weaponized cyber tools.

Back in 1945 no one demanded of the nuclear scientists a complete assessment of the effects of the gamma ray output, responsible for the generation of the three types of pulses[26] that produce the nuclear EMP phenomenon, because these effects were not completely understood by the nuclear scientists at that time. They knew about the EMP but they didn't completely know about the magnitude and significance of EMP effects. We know this because historical records show that even though the electronic equipment used during nuclear tests in 1945 was shielded for protection, the equipment was negatively affected, in spite of the protective shielding. Seven years later, during British nuclear testing, they also experienced instrumentation failure due to EMP. Furthermore, during the Starfish Prime high altitude nuclear test, nuclear scientists realized that the effects were much larger that they have expected, causing damaging effects on electrical and electronic infrastructure in Hawaii.

Why the double standard? Why do we want to have perfect assurance that nothing will go wrong with collateral damage when releasing a weaponized cyber tool? In the delivery of every single weapon collateral damage is both unavoidable and unpredictable. The dissemination of cyber code throughout the Internet is a very precise science, and we do have means to determine the cyber area of exposure of any such particular tool. Do we know how to define the area of coverage of a cyber tool? Yes, absolutely. However, this is not the issue. We cyber professional know how to do it, but we cannot control the errors in configuration that any network administrator might have introduced into his network AOR. There are imponderable factors, today as they were yesterday. After all the calculations were completed and revised concerning the coordinates for the precise point of impact of Little Boy and Fat Man, did we achieve perfect accuracy on their delivery? The

---

[26] E1, E2, and E3, as defined by the International Electrotechnical Commission (IEC)

answer is no, because of imponderable factors. So, why do we want a 100% warranty in all details associated with the delivery of weaponized cyber tools? And why do we insist now in holding this unrealistic expectation that we didn't hold before? Furthermore, how do we explain all the quantifiable parameters we have at hand for a responsible delivery of such tool to the legislators, judicial monitors and policy-makers, if they are not prepared to understand the unique cyber language and lexicon, because they have shown unwillingness to learn it?

There are times in history when we have to abandon the way of thinking that is familiar to us, in order to successfully and productively embrace our awakening to a new reality. During the 15th century we gave a unenthusiastic welcome to the heliocentric model discovered and sponsored by Copernicus, because it didn't fit into our familiar geocentric way of thinking. Even worse, two centuries later we were labeling and convicting Galileo as a heretic, and forced him publicly to recant his endorsement and advancement of the heliocentric solar system model. It took three centuries for those who convicted Galileo heliocentric views to rescind their opposition. Today we don't have three centuries to afford ourselves a comfortable transition from our familiar historical warfare doctrines into the new and unprecedented cyber conflict paradigm. The ever present threat of a cyber conflict requires that we take the crucial step of adopting a new mindset to deal with cyber affairs, and to abandon the futile effort of trying to force the entity of cyber conflict into the old models of traditional warfare.

The effects of any cyber operation may remain unwitnessed to the average population, but they are no less tangible. This cognition is not a requirement for the definition of what constitute a cyber conflict, an event that may occur with or without human awareness. The best illustration for this principle is the examination and analysis of the timeline of any cyber intrusion, which may span several day or weeks, and in some cases even months, between the moment of the launching of the cyber intrusion and the discovery of such an intrusion. Perhaps the most relevant case at this point is the Stuxnet cyber intrusions, which according to qualified forensic evidence, we now know that the time span between the intrusion and the discovery took over several months. The public disclosure of the Stuxnet infection was originally reported by VirusBlokAda on 17 June, 2010, but the impact of the cyber infection did not begin to gain recognition until late June. Siemens did not publicly acknowledge the impact of Stuxnet until 14 July of that year. However, the question remaining is: how long was since the Stuxnet infection was planted on the affected systems? The time stamp forensically recovered from some of the main Stuxnet components revealed that such files were compiled as early as January 2009. Consequently, we can only retroactively establish a timeline for the Stuxnet infection that spans over a year, and yet, we remained publicly unaware of it for over a year.

The fact that a cyber event does not requires human witnessing is perhaps one of the main hurdles in defining and codifying the presence and scope of a cyber conflict. This conundrum belongs to the sphere of activity where only qualified cyber SMEs can and should lead public awareness, and provide especial guidance to the human elements comprising the policy-making and legislative bodies. These two bodies must seek and achieve a balanced synergy in becoming the seminal foundation for a proper codification of cyber conflict legislation.

Let us consider the cyber legislative scene at the closing of 2011. As of November 2011 the CSIS[27] reported in their blog[28] the undertaking of a government proposal for legislation on dealing with cyber criminal activities, and securing the national critical infrastructure, in addition to enhancing the protection of networks. The blog adds that the release of this proposal occurred on May 12, 2011, and is still undergoing discussion in several House and Senate Committee meetings. The question that is relevant at this point is: has there been any consultation with cyber SMEs during these six months of discussions? Furthermore, is there any awareness among the members of the involved committees regarding what are the best practices and the main technical issues involved in securing the national critical infrastructure?

---

[27]   The Center for Strategic and International Studies
[28]   http://csis.org/blog/cyber-legislation, Nov 10, 2011

These are not rhetorical questions. As this author has already pointed out in this book, there is very scarce consultation between policy makers and certified and qualified cyber SMEs. Another aggravating factor in codifying relevant cyber legislation is the nebulous view of the technical aspects involved in securing the complex industrial control system (ICS).[29] Legislation on this very critical entity without consultation with qualified cyber SMEs has the potential to introduce even more vulnerabilities, by focusing exclusively on ICS aspects we consider more critical than others. I personally participated in a meeting with leadership of an important organization, and I outlined for them the criticality of the threats to the ICS operating in their AOR, and I even provided a realistic scenario to highlight the level of criticality. The person responsible for the operational availability of their ICS responded that the scenario I presented was feasible, but his main concern resided in maintaining functionality of a more tangible and immediate reality, such as keeping running water and functioning toilets. No one will deny the importance of these domestic necessities, but it is an error to assign paramount importance to these two, to the detriment of other equally or more critical ICS components, simply because we are not aware of their criticality. Consultation with qualified cyber SMEs will introduce a more balanced view of ICS, leading to a professionally prepared priority list of the ICS components.

This very same cited legislation proposal highlights the utilization of intrusion prevention systems (IPS), thus exhibiting the widespread gullibility in autonomous and automated systems that promise what they cannot possibly deliver, simply because the goal is unattainable. IPS systems began to appear in the market after the decline of their predecessors, intrusion detection systems (IDS). The IPS appliances experienced a remarkable market success because of the fallacies upon which the marketing strategy is based. The IDS were portrayed as representing a passive defense model, because they detect and announce a cyber threat, but they do not take any action to prevent or mitigate the detected threat. Consequently, the IPS marketing strategy was launched with the promise that the implementation of IPS appliances would allow the

---

[29] See chapter 7 ahead

users to transition into an active defense model, whereby the IPS appliances would detect and stop a cyber threat, thus preventing any cyber intrusion from taking a hold into the enterprise network assigned to defend. Does this sound too good to be true? It does, because is not true.

Let's explore the issue of a cyber threat in the context of the average enterprise that lacks personnel certified, qualified and experienced as cyber defenders. The marketing strategy used by IPS vendors became the answer to all the fears of the great majority of customers that lack understanding on the dynamic of a cyber threat. The great majority of both commercial and governmental organizations lack the qualified cyber professionals capable of defending against a cyber threat. The majority of the unqualified personnel assigned to the task of cyber defense are simply responsible for documenting and reporting a cyber threat or intrusion. They are not capable of understanding the cyber threat dynamic, and they simply "react" by following a script instructing them the required steps to report a cyber incident. Under these conditions the IDS model can easily be made the escape goat, and portrayed as a "passive" and inefficient system. This is a deflective mechanism, designed to shift the responsibility from the human assets responsible for dealing with the threat, and placing the blame on the IDS model. At this point it's necessary to reiterate the main premise of this book:

A cyber conflict is a battle of brains, not of machines

The concept of IPS is a positive addition to the paradigm of defense in depth, when, and only when, it becomes another layer of defense, but it becomes a fallacy and a source of false sense of security when it is implemented as the only means of cyber defense. Vendors' advertisement misrepresents the capabilities of IPS appliances by stating they are better than their IDS counterpart because an IPS is designed to be connected in-line, and thus able to actively prevent intrusions by blocking the malicious activity. This author has personally worked with IDS, and they are also designed to be connected inline, and implement active blocking against malicious cyber activities. Most organizations opt not to connect either IDS or IPS appliances inline because they became a single point of failure on network availability. If anyone of these two systems are connected inline, and they experience an operational failure, either one will disrupt network connectivity for the rest of the enterprise network connectivity. See diagram below for an illustration of this issue, where the IPS is connected inline, offering the capability of blocking traffic directly at the IPS level, but creating a single point of failure at the same time.

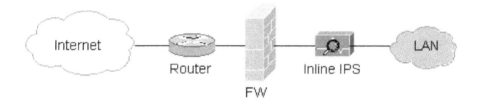

Traditional IDS appliances are customarily in an out-of-band connectivity model, in order to offer unrestricted monitoring of all traffic inbound and outbound of the LAN, while avoiding the single point of failure problem, as illustrated in the below diagram.

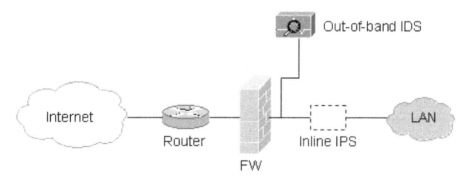

However, the main issue in discussing this particular issue regarding the value of IDS and IPS appliances does not center on these systems themselves, but on the purpose for which we deploy them. This author exposes the fallacy of relying on these systems alone because by an in themselves they do not represent the solution to cyber defense. Both cyber attack and cyber defense depend exclusively on the technical cyber knowledge, shrewdness and expertise of the two sides involved in a cyber conflict. Within this context IDS and IPS systems represent only a subordinate tool, under the supervision and final decision of qualified cyber SMEs. The preponderance of these systems, elevated to an autonomous role that it's alien to their design, is the result of decisions made by uninformed leaders, from both commercial and governmental organizations, who think that machines can actually provide protection to a network, while excluding the role of cyber SMEs.

And why would they decide to exclude the cyber SMEs from the cyber defense equation? Some took this decision because of a faulty economic assessment that led them to believe that the "Goddess Automation" would save them from the expenses of employing cyber SMEs. After all, there are so many other industries flourishing by adopting automation, so why not applying the same solution to cyber defense, and save all the money previously allocated for cyber SMEs salaries? Others adopted the decision of automation because they didn't have a choice: they had no qualified cyber SMEs to do the job of providing cyber defense.

Let us summarize what we have just stated:

The addition of IDS and IPS to the paradigm of cyber defense in depth is a positive factor, when, and only when, it becomes another layer of defense, subordinate to the leadership of qualified cyber SMEs.

The only real difference between IDS and IPS is that the latter attempts to establish a network behavior baseline, and attempts to block network traffic than deviates from the learned baseline. However, the learned baseline network behavior is limited by the allotted time dedicated to learn such behavior, and that time is never long enough for the IPS to "learn" every possible "normal" and "authorized" behavior traversing the monitored network. Thus, the false positives generated by IPS become an additional administrative burden for network administrators.

While assigned to the task of protecting a government network, this author was invited by a vendor to test an IPS recently purchased and installed in this network. After the brief network behavior learning period allocated for the IPS, the number of false positives generated by the IPS exceeded the potential benefits of its use, and it was discarded, after the vendor was unable to implement a configuration capable of rendering false positives within a manageable and acceptable range. The empirical assessment leading to this decision was possible because the cyber network defense task was performed and implemented by a knowledgeable and experienced team of cyber SMEs.

The perfect balance on the cyber defense paradigm is achieved only when it is established on the foundation of qualified and experienced cyber SMEs, who understand the principles and dynamics involved on a cyber conflict. Automated cyber defense tools become a balanced complement to the knowledge and tactics of such SMEs, and only when implemented in a supportive and subordinate role to the same SMEs. The following diagram illustrates this balanced concept of cyber defense in depth, attainable only when the cyber defense paradigm is based on the presence and tactical leadership of cyber SMEs, assisted by any number of available combinations of automated cyber detection systems.

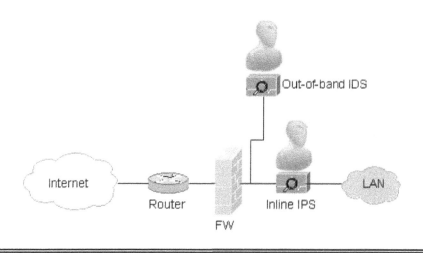

A cyber conflict is in its very essence a battle of brains, developing in the battlefield defined by the binary code environment where the offensive and defensive actions clash, and where only the superior knowledge on networking principles and binary code processing will determine the victor. The side choosing to take initiatives will always have the greatest potential for victory, and the side choosing a reactive mode of defense is condemned to a certain defeat. This latter is the option that most developed nations have taken, by throwing all their resources into automated equipment dedicated to emulate a formidable cyber Maginot line in the sand, while discarding the role of the human brain in the defensive equation, erroneously assuming that a cyber conflict is fought by cyber hardware and software, and relegating the human element to a role of spectator. This fallacy is fed by the lack of knowledge on how a cyber conflict is fought and won.

One simple illustration on the role of the unawareness fueling this fallacy can be drawn from the daily life of an average individual computer user. When confronted with the reality of becoming a victim of a spear phishing attack, for example, this typical user tends to wonder why the attack succeeded in spite of the presence of an anti-virus (AV) software product installed on the victimized system. This surprised user doesn't understand that the two issues are completely dissociated. AV software is not designed to protect against spear phishing, simply because this exploit technique relies on the user being lured into taking an action that initiates the exploit process, thus making the user the active agent responsible for bringing a cyber exploitation into one's cyber system. AV code is designed to recognize the malicious cyber signature of a known malware. Conversely, the AV code is not designed to prevent the user from initiating an action that will activate a cyber exploit.

There is a vast variety of cyber exploits, and only a small portion of these exploits arrive in the form of a malicious code corresponding to the known signature of computer malware. The presence of an

AV is not a guarantee of total protection against all cyber exploits seeking for unaware victims traveling in the cyber spectrum. The cyber jungle is dangerous, and unaware explorers on any scale, individuals or enterprises, will fall prey to unscrupulous cyber predators if they lack cyber awareness.

Chapter 6. Asymmetric cyber conflict

Complete cyber security is unattainable, but cyber defense is feasible to the extent that the defending entity is willing and capable of deploying and implementing the proper cyber defense layers of security. At this level military and economic power plays an insignificant role, if any at all. Only knowledge of cyber spectrum is the determining factor. Those who wield in-depth knowledge of the cyber spectrum will always win, and those who simply talk about cyber spectrum issues but lack in-depth knowledge will always lose. Knowing where binary vulnerabilities reside and how to exploit them for one's own agenda is the sure path to victory in any cyber conflict. This is where cyber knowledge becomes the ultimate equalizer in a cyber conflict. The tremendous disparity between those who know and those who don't know generates the concept of asymmetric cyber conflict.

This author entertains the discussion of asymmetric cyber conflict simply as a concession to the now popular use of this terminology. When the term was coined[30] it was designed to analyze the puzzling reality of a powerful nation losing small wars against another nation with a remarkable disadvantage in material power. How can the victory of the "weak" against the "strong" be explained? This is the topic under study by the adherents to the asymmetric warfare doctrine. However, there is no asymmetry when it comes to a cyber conflict, because material power is not a determining factor in the cyber spectrum, and this is the entire premise of this book. Cyber prowess is the ultimate equalizer, and cyber prowess depends primarily and ultimately on brain power and applied knowledge in cyber principles. Consequently, there is

---

[30] "Why Big Nations Lose Small Wars: The Politics of Asymmetric Conflict," Andrew J.R. Mack, World Politics, Vol. 27, No. 2 (January 1975),

no asymmetry in the cyber spectrum, because cyber knowledge is available to every single human being with the capacity to study and learn. Therefore, in cyber there are no "weaker" nations, except for those who opt not to invest time and effort in learning the cyber principles. And that is not an imposed weakness; that is a choice.

Cyber defense is impossible from a Maginot doctrine point of view. The Maginot Line (French: Ligne Maginot) was a line of concrete fortifications, tank obstacles, artillery casemates, machine gun posts, and other miscellaneous defenses constructed along the borders of France with Germany, based on the historical tradition of relying on the success of static defenses, and the World War I doctrine of static defensive combat model. This fortification system was strategically and tactically ineffective, as the Germans invaded Belgium and flanked the Maginot Line, thus nullifying its purpose.

There is a cacophony of opinions—and we all know about opinions—tossing around the term "cyber war", and those behind such opinions are not speaking from a technical point of view, but rather from a particular agenda they must sustain, considering their current or former position. Perhaps one of the most quoted statements representing a misled concept on cyber spectrum recently appeared on an article, voicing the opinion that "after land, sea, air and space, warfare has entered the fifth domain: cyberspace"[31]. Cyber spectrum is not a domain, but a ubiquitous sphere of cyber existence, global and borderless, encompassing all other traditional domains of warfare, establishing synergetic dynamics with all of them, but never fusing with the other domains, because cyber always retains its unique essence. Cyber exists and interacts only within the sphere defined by cyber binary code. This lack of understanding on the true nature of the cyber spectrum causes a rippling effect, generated by ignorance, on a corollary discussion regarding the current status of cyber warfare. Some proclaim that cyber war has already started and the US is already losing it, while others retort that there is no cyber war.

---

[31] http://www/economist.com/node/16478792

Cyber conflicts are ubiquitous, and this reality does not depend at all on whether or not legislation and semantics allow for the usage of the term cyber warfare, thus rendering this term empirically irrelevant.

The asymmetry in a cyber conflict springs from the disparity between the opposing sides, one knowledgeable in the cyber techniques, and the other lacking in-depth knowledge of the scope and techniques germane to the cyber spectrum. However, knowledge is not the only factor, but rather the foundational one, on the asymmetric cyber conflict equation. Adherence to established RoE, or the complete disregard for such RoE is also a very important contributing factor fueling the asymmetric cyber conflict dynamic. When a particular cyber conflict participant manages to combined both the knowledge factor and a total disregard for RoE, the asymmetry becomes critical and virtually impossible to manage.

Let's consider the asymmetrical factor that is being forged in our own present time. The secure internet protocol known as IPv6 has been deployed in virtually all the world continents, with the exception of North America at large, and specifically among the United States government various organizations. There are five Regional Internet Registry (RIR) organizations administering the distribution of IP address blocks, shown in the following table:

RIR name	Global area
AFRINIC	African region
ARIN	Canada, some Caribbean regions, and USA
APNIC	Asia, Australia, regional neighboring countries
LACNIC	Latin America, and balance of Caribbean regions
RIPE	Europe, Middle East, Central Asia

When tabulating the number of allocations for IPv6 address blocks, RIPE is leading, showing a remarkable growth in 2008 and 2009. Recently, ARIN allocations also increased to the point of showing a slight increase over APNIC. LACNIC and AFRINIC have comparatively fewer allocations. The global allocation of IPv6 address blocks at the end of 2009 is shown below:

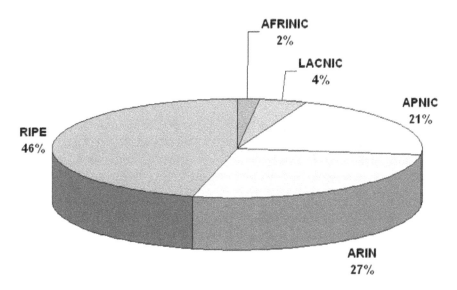

The users with a more extensive use of IPv6 will gain an increasing knowledge base and experience in using this advanced network protocol, while those who continue delaying the adoption and deployment of IPv6 at a large scale will remain asymmetrically inferior to those who are already using it. During a near future cyber conflict, where IPv6 is the defining network environment, we in North America have already become losers, because we lack the empirical knowledge on IPv6. Even during a theoretical absence of cyber conflicts, we have already placed ourselves at a tremendous disadvantage, since we cannot operate and much less dominate on a network environment globally governed by the network principles, capabilities and security factors built into IPv6.

It is undeniable that the technical transition from IPv4 to IPv6 presents a very complex challenge, but delaying the transition will only increase the complexity factor. All three feasible mechanisms devised for this transition require access to IPv4 addresses, and if we wait until all IPv4 address blocks are exhausted[32], then the

---

[32]   Exhaustion date is optimistically estimated as 2012

transition mechanisms themselves become unfeasible. These transition mechanisms are known as dual-stack, tunneling, and translation. When a network device is capable of implementing both IPv4 and IPv6 network layer stacks, it is operating under the "dual-stack" transition mechanism, enabling the co-existence of both IPv4 and IPv6. When isolated IPv6 network devices need to connect to other IPv6 devices, they must do so over an intermediate IPv4 network infrastructure, thus requiring a tunneling mechanism to carry IPv6 network traffic over an IPv4 network environment. Finally, when IPv6-only network devices need to connect to IPv4-only network devices, an intermediate translation device is required to implement the required translation between these two dissimilar network environments. Thus, all these three mechanisms, dual-stack, tunneling and translation, require access to IPv4 addresses, therefore dictating that these transition mechanism must be implemented before the IPv4 addresses are exhausted.

Multiple companies took the initiative on an early start in transitioning into IPv6, and the extensive experience they have accumulated in doing so places them in an advantageous security posture. An Asian company[33] pioneering in the field of IPv6 deployment and solutions boasts that as the leader in this field, they started providing full-scale IPv6 operations in 2001, and in 2003 completed the implementation of IPv6/IPv4 dual-stack technology on the backbone of national and international networks. A reputable study[34] endorsed by 30 governments[35] reports that by early 2010, IPv6 constituted only a small portion of the Internet

---

[33] NTT Communications as the international and long distance service arm of NTT (Nippon Telegraph and Telephone Corporation)

[34] The Organization for Economic Co-operation and Development (OECD) recently published report on "Internet Addressing: Measuring Deployment of IPv6", April 2010.

[35] OECD membership includes Australia, Austria, Belgium, Canada, Czech Republic, Denmark, Finland, France, Germany, Greece, Hungary, Iceland, Ireland, Italy, Japan, Korea, Luxembourg, Mexico, the Netherlands, New Zealand, Norway, Poland, Portugal, Slovak Republic, Spain, Sweden, Switzerland, Turkey, UK, and USA

infrastructure, and only 0.16% of the top 1 million websites had an IPv6-enabled website in March 2010.[36]

This asymmetrical condition is not imposed upon us, but is rather a self-inflicted disadvantage. Since in cyber spectrum knowledge is the equalizer, ignorance and lack of experience is a dreadful destabilizer.

The asymmetry in cyber conflict is further enhanced by an irreversible economical model affecting industrialized nations in general and USA in particular. Such model is globalization, the modern process integrating regional and national economies through a global network of communication, transportation, capital flows, technological dissemination, and trade. Since the inception of globalization until now it is virtually impossible to find a single national entity still manufacturing native products and components used in cyber devices and cyber network systems. Globalization became the template to maximize profits by diversifying manufacturing process and building cyber components throughout the entire global village, seeking to maximize the profits by fabricating components under the most advantageous financial conditions, regardless of socio-political boundaries. This economical model gave birth to the supply chain threat, one of the most damaging and virtually unsolvable issues in cyber security. This supply chain threat is both and attack vector and a victim itself.

The opinion voiced in innumerable press announcements and organizational reports indicating that the U.S. is facing "thousands of attacks daily" is a gross misrepresentation of the root cause, and a pusillanimous attempt to deflect responsibility. Cyber attacks target vulnerabilities, and cyber vulnerabilities are neither regional nor national. Cyber vulnerabilities are global, and the result of inherent weakness introduced in the binary code of both operating systems (OS) and applications (apps). The main contributing factor in these inherent weaknesses is the motivation to produce binary code that satisfies the quest for an increase demand on multiple features that are both entertaining and user friendly. Binary code

---

[36]   OECD report, pp. 4, 5

security is often sacrificed in the altar of marketing and profitability. These inherent weaknesses give birth to the unceasing demand for patches, as the vulnerabilities are discovered and exploited. When any given nation thinks of itself as being in the cross-hair of cyber attackers is because the same nation has the most vulnerabilities accumulated in their installed-base of OSes and apps.

Cyber attacks are systematically based on the discovery of binary code weakness, thus all cyber attacks are directed at these weaknesses, not at any nation in particular. But when a nation distinguishes itself for the highest index of vulnerabilities, then such nation may perceive itself as a victim, thus deflecting the responsibility for researching, detecting and correcting the binary code weaknesses. In any given cyber conflict there are no innocent victims; there are simply targets of opportunity. If a nation chooses to offer itself as a target of opportunity, by ignoring the responsibility of protecting itself by correcting the binary code weaknesses, then the cyber wounds suffered during a cyber conflict are self-inflicted. Society and legislation at large have not yet begun to comprehend the concept of accountability associated with operating cyber systems and applications. We have a sizeable legislation regulating the use of transportation devices, and the corresponding licensing demands required for operators of transportation devices, but for an exponentially more powerful and potentially damaging cyber device we require no accountability whatsoever. It's always so much easier to play the victim and blame it on the nefarious attacker that found and preyed on our cyber weaknesses. We prefer to pretend that the Darwinian axiom of the survival of the fittest doesn't apply in the cyber spectrum, and yet we ignore that it is precisely in this cyber spectrum where this axiom finds its ultimate fulfillment.

There is an important question to consider when dealing with the concept of asymmetry in the cyber arena. What is the role of the individual, organization, or nation in presenting themselves as a target? Is a driver responsible for the mechanical and electronic maintenance of the driver's vehicle? Is the driver responsible for the damage the driving vehicle may inflict on human life or property due to a malfunction known to the driver? The legislation written in regards of transportation accountability holds the driver responsible for the damage caused by the driving vehicle, but there is no legislation or accountability for the damages an irresponsible user of a cyber system incurs due to negligent use of the malfunctioning system. Is the concept of cyber asymmetry standing on the grounds of a mythical superb and superior adversary that is capable of exceeding and nullifying all the defenses of the target system, or is this cyber asymmetry the result of a willful negligence in operating and maintaining a cyber system that is lacking all required defenses?

There is a myriad of individuals, organizations and nations hiding behind this fallacy of cyber asymmetry, a term coined as a shameless attempt to avoid accountability and responsibility when operating cyber systems. Any cyber system is essentially based on cyber code, and cyber code is written by human beings with an agenda in mind, most of the time driven by a natural and legitimate desire to achieve functionality and profit. As such, the cyber code will always include weaknesses that are the result of design oversight, or a risk that is considerable acceptable in order to obtain an intended profit margin. An adversary studying a particular cyber code will eventually discover such weaknesses, and may decide to exploit them. This is an inevitable risk in the cyber sphere.

The inevitability of weaknesses in cyber code is an issue we must face with responsibility and accountability, by designing and maintaining a cycle of security upgrades responsibly and timely applied to the vulnerable code. These security upgrades are built and disseminated free of charge, to all users affected by the code vulnerability in the case of cyber applications in general, and anti-virus (AV) code in particular. This later case is not based on a code weakness, but on the fact that AV require a constant updating of the cyber signatures responsible for recognizing and nullifying the threat pose by the constant flow of new malware. There is always a window of opportunity when the vulnerable code is susceptible to a cyber intrusion, a window defined by the time span between the discovery of the vulnerability and the dissemination of the corresponding security upgrade. This window is known as a zero-day exploit period. During this zero-day period any vulnerable code is subject to an unavoidable cyber exploitation. However, this condition of exploitability ceases to exist the moment the security upgrade becomes available from the vendor, or when a work-around solution is made available by third-party sources who have found an interim solution until the security upgrade becomes available. This safety mechanism works extremely well, but it's persistently ignored by individuals, organizations, and nations. This author has collected data from large organizations, extracted from system upgrade logs, showing that such organizations delayed the application of available patches for periods as long as 90 days, and in a few cases, even longer.

The bulk of the major exploits affecting large organizations, both in the commercial and the government arena, are due to negligence in applying the available security patches in a timely manner. This irresponsible behavior is persistently disguised by attributing the exploit to "sophisticated adversaries" who are misrepresented as being so advanced in cyber matters as to be able to defeat any cyber defenses. This is a shameless cop-out mechanism, used and abused by commercial and government organizations, to avoid accountability for their negligent behavior in delaying or ignoring the security upgrades made available to protect against cyber exploits. "*A single negligent user or unpatched computer is enough to give attackers a beachhead into an organization from which to mount additional attacks on the enterprise from within, often using the credentials of the compromised user*".[37] The abused characterization of adversaries as "sophisticated" has become the most popular deflecting mechanism to avoid the issue of accountability. As recently as 2010 there were national leaders using this terminology when testifying before Congress in order to explain the severe threat assaulting U.S. information systems,

---

[37]   Symantec Internet Security Threat Report, Trends for 2010, Volume 16, published April 2011, p 8

and the escalation of adversarial cyber activity characterized as one of "extraordinary sophistication."[38]

This type of assessment is intended to be used as a political statement designed to deflect the responsibility for the escalation on cyber threats. An assessment from a group of qualified cyber SMEs will show a remarkably different picture; our current condition of cyber insecurity is the direct result of neglecting to take the responsibility for securing our information systems with the assistance of qualified cyber SMEs, armed with the experience and the knowledge to design, deploy and maintain a proactive cyber defense posture. The adversaries are not sophisticated; they are simply persistent in their nefarious task, and when an individual, organization or state neglects to establish a proactive cyber defense mechanism, they actually become a target of opportunity and actively invite the adversary to mount an attack. In the opinion of this author, with over 15 years of experience in dealing with cyber security and cyber attacks, approximately 90% of all cyber attacks are based on either patched or preventable vulnerabilities.

There is no perfect cyber defense, and there is no perfect cyber attack either. The difference between victory and defeat in a cyber conflict resides in the wise and balanced decision between what we can defend and what we can concede to the adversary. But the true key opening the path to victory is the capability to deal with a cyber conflict in a nonlinear manner. The cyber spectrum is so complex and dynamic that the two conflicting parties have to enter into a demanding interaction with a phenomenon that is always greater than the sum of all the parts perceived during the conflict. The interacting elements in the conflict are continually evolving in a dynamic that pertains to the realm of the chaos theory.

There are instances during the conflict when nothing seems to make sense, but this is the result of the erroneous perception

---

[38] Testimony before the House Permanent Select Committee on Intelligence, Annual Threat Assessment, 111th Congress, 1st sess., 2009.

of seeing the cyber conflict as a linear progression, where we expect a single cause-effect dynamic. In the complexity of the nonlinearity nature of the cyber spectrum we may have multiple, and seemingly dissociated factors interacting and influencing one another, and a linear analysis of cause-effect may very likely lead to the erroneous conclusion. This is not to say that a correct analysis of such complex entity is unattainable; rather, this is to say that we need to analyze the cyber conflict at hand from a nonlinear perspective.

Is there any way to overcome the current self-imposed asymmetry in our country? Do we have any human resources capable of bringing balance into the cyber conflict by nullifying the current asymmetry? The human resources are indeed readily available within the pool of cyber experts in our nation, and within this cyber pool there are very well qualified and extremely talented individuals capable of great accomplishments, and equally capable of generating a steady influx of cyber solutions to dispel the current asymmetry. Within this pool we have individuals with extraordinary cyber knowledge and cyber expertise, but they are not necessarily aligned within the parameters required by our government's RoE. Many of these extremely gifted cyber assets are neither prepared nor willing to work within the constraints of such RoEs. Many of these assets align themselves with the libertarian[39] philosophy, holding to the unrealistic view that every individual is entitled to know about everything they want to know, and express themselves in every topic that may capture their interest.

A clear illustration of this libertarian attitude occurred during the recent Black Hat Conference 2011 in Las Vegas, where this author was in attendance. The topic under consideration was presented through the opinion of three panelists, who were selected for their relevant expertise in current examples of hacktivism. The panelists offered their opinions in response to questions and

---

[39] Individuals who are tenacious advocates of the concept of civil liberty, placing a great emphasis in the individualistic expectation of unimpeded expression of thought or conduct.

opinions presented by the topic moderator. One of the panelists was incognito, wearing a dark and large scarf covering his entire head, except for a slit in front of his eyes. This panelist, after presenting his opinion on the cyber entity "Anonymous", was criticized by a member in the audience, who challenged the panelist with the comment: "Why should I believe what you are saying when you are hiding behind a mask and you don't show yourself to me?" This challenge exhibits the very essence of the logical flaw in the minds of those who align themselves as libertarians. This challenge originates in the arrogant attitude of considering that the individual holds the criteria to discriminate between what is credible and what is not, and the single criterion springs from this unique demand: "unless you fully disclose your identity before me I will not grant any credence to what you are saying. I demand to see your face before I can even begin to consider the credibility of the content of your speech."

The fundamental philosophical flaw in this libertarian attitude is the arrogance of believing that one is the very authority to judge other peoples opinion. Truth is discernible by those who are able to assess and identify the markers of a credible statement. Within the human sphere of cognition truth is always a relative entity, and this as an inescapable corollary of our limited cognition and lack of absolute knowledge. Thus, within the human sphere of cognition truth is a verbal or written expression of a reality that is attainable within a limited historical context[40] and supported by witnesses cognizant of the constituting elements of such truth. Those in touch with the defined historical context can attain cognitive knowledge of the elements constituting the expression of truth, and therefore are enabled to discern the accuracy of such expression. This definition of truth does not require knowledge of the personal identity or facial recognition of the person uttering the expression of truth. Truth is self-evident within the context of this definition of truth.

The obvious flaw in the faulty logic of the challenger, demanding to have facial recognition of the person uttering the statement

---

[40] Author's personal definition of a historical truth.

during the discussion, is quite evident on the outcome of this event. After consulting with the entire audience (about 300 people), the moderator of the session discovered through a vote of hands that there were less than 10 people who wanted the masked spokesperson to reveal his facial features. However, the masked individual volunteered to reveal his countenance, and after doing so, there were only three individuals who recognized his identity, thus proving that the credence ascribed to his statements was not based on facial recognition, but on the contextual acquiescence that the audience was granting to the arguments presented by the unveiled spokesperson. This is the libertarian challenge we must face if we are to increase the contingent of knowledgeable, gifted, and experienced cyber experts into the fold of defenders of our national cyber assets. They may be very capable at the technical level, and very proficient in adding a tremendous arsenal of cyber acuity into our cyber defensive program, but they need to reassess their priorities on how, why, and when to use their knowledge, within the lawful and ethical boundaries we accept in order to provide an enhanced cyber security posture for our nation.

In cyber there is a very fine line between what is right and wrong, ethical and unethical, lawful and unlawful, and those who are willing to serve as qualified cyber warriors must be willing to make a commitment to live by an accepted and endorsed code of ethics. Cyber knowledge is ethically neutral, but the application of such knowledge must follow the RoE set approved and endorse by a legitimate authority.

Life in cyberspace is not a party; it's a battle for survival, and in the cyber jungle knowledge of the terrain, constant vigilance, avoidance of unnecessary risks, and respect for the rules of cyberspace means life. Disregard for all these factors means unavoidable cyber death. And those who are dying or losing the battle in cyberspace every day are doing so because of their own negligence in living according to the rules of the cyber jungle. Claiming ignorance of these rules will not save anyone from suffering cyber death. In the cyber jungle knowledge is not only power; knowledge is also life.

Excursus: The myth of classified cyber information

Many government organizations operate under the erroneous conviction that only classified information on a cyber issue or conflict is worth their attention. In doing so they make two simultaneous mistakes: One, they delay the collection and the analysis of the relevant cyber information until is introduced into classified networks, because only then government organizations consider the collected information to be properly "vetted" and credible. Two, they intentionally over-classified the collected information in order to create the illusion that the data is indeed vetted, and more credible than the corresponding data already available in open sources. The sad reality is that the very same general data corresponding to a cyber issue or conflict has slowly permeated through several layers of situational awareness (SA) in the government ranks, and the very same data available on open sources has finally been captured and regurgitated on classified networks. This process of assimilation of cyber data introduces an unnecessary delay that increases the self-induced asymmetry when dealing with a cyber conflict, and creates a false sense of security by pretending that the cyber data finally collected on a classified network is more reliable than the counterpart available in open sources, thus creating a delay of weeks and even months before appearing on classified networks.

The faulty premise underlying this erroneous conviction is another expression of crass ignorance regarding the nature and essence of the cyber spectrum. By definition the cyber spectrum is cyber code, and this cyber code exists in open sources from its very inception. Therefore, any cyber issue or conflict will always be based on a particular cyber code issue, and thus, it will always be unveiled and disclosed on open sources first. The only possible exception to this rule of thumb it would be a piece of cyber code that has been developed from its very inception under the auspices of a government organization and created from the beginning inside a classified government network. This type of scenario is very rare.

Let's take the case of a zero-day exploit weaponized to launch an offensive cyber action developing into a cyber conflict. The offensive party will research a particular application cyber code, and when a weakness within the selected code is discovered, a zero-day exploit is born. The attacker will research what organizations are using this particular vulnerable application, and will launch the cyber attack by deploying the recently found cyber exploit. Or conversely, the attacker can simply research the enormous amount of vulnerabilities found by neutral researches, take their findings, test the newly discovered exploit, and launch it when proved to be a deployable vulnerability. Many new vulnerabilities are disclosed during cyber conferences around the world, and attackers are continually capitalizing on this course of events.

Before the reader falls into the temptation of expressing despair for the terrible world in which we live, let's consider the self-induced asymmetry present in the zero-day scenario under scrutiny. Why are we assuming that the attacker always has the advantage? Why are we not, on the defensive side of the fence, investing the time and effort of maintaining our SA on zero-day exploits as the attackers do? Remember the title of this book, "The Cyber Equalizer". Whatever tools are available to the attacker are also available to the defender. But if the attacker takes the initiative of investing resources in maintaining SA on cyber exploits, why do the defenders are always lamenting how terrible is living under the constant threat of cyber attacks? Both sides are entitled to obtain the same information at the same time, but if one side chooses to maintain SA while the other side languishes in despair and inactivity, always waiting for the next blow threatening their networks, in a viciously self-induced asymmetry in both SA and initiative, then let's not complain about the terrible predicament in which we live, caused by our own defeatist inertia.

A zero-day exploit should not be considered a doom day event. The professional cyber community maintains a very dynamic exchange on cyber vulnerabilities research throughout numerous and reputable cyber fora. Any responsible and proactive cyber defender can and should gain access to this data, find the corresponding set of data applicable to the defending network enterprise, and designed and implement a preventive course of action. This proactive and preventive initiative will provide an acceptable degree of cyber security against the emerging zero-day exploits, via the implementation and application of interim security mitigation techniques, until the security update is released by the threatened application's vendor. This author has successfully taken this course of actions on many occasions, with satisfying results. Of course, in order to obtain these satisfying results an organization must be willing to dedicate human and technical resources to maintain SA on the never ending sequence of emerging zero-day exploits. How can any organization benefit from this proactive course of action model? By employing certified and experienced cyber SMEs that understand the essence of cyber, know how to find, collect and analyze the cyber data pertaining to emerging cyber vulnerabilities, and design and implement a mitigation technique tailored to the defended network.

This is the walking on the dromonikian road! This is the path of the true cyber warrior, previously introduced in chapter 4. This road to victory is not accessible by means of depending exclusively on automated cyber security devices in conjunction with a contingent of metamorphosed operators. Stepping on the dromonikian road requires in-depth understanding of fundamental cyber security principles. When we lack this understanding, we claim false victories and we wrap ourselves into a false sense of security. This author recently read a cyber incident report claiming a successful defense implemented by an organization targeted with a DDoS. The report stated that the DDoS attack was unsuccessful thanks to the quick reaction of the defending organization by expeditiously creating and deploying tailored AV signatures to defeat the DDoS. This is not only an impossible claim, but it is leading the targeted organization, and any other enterprise sharing their report, into a false sense of security.

A DDoS is a network attack that targets and depletes the memory resources of a cyber system to the level where such system can no longer operate, thus the name of the attack, "denial of service". The service that the targeted system is designed to offer is no longer available because the memory of the said system has been exhausted, thus rendering the system non-operational. The target of this attack is the system memory. In this type of attack there is no malware code involved, and the system can recover after rebooting it.

On the other hand, a malware attack is based on the presence of a malicious code that it is injected into a particular application hosted on the targeted system. The malware code attaches itself to the targeted application and maliciously modifies the behavior of such application, in accordance to the attacker's agenda and purpose of the cyber attack. The malware victim cannot recover by simply rebooting it. It requires sanitization (best case scenario), or a system rebuild (worse case scenario).

Consequently, it is absolutely impossible to claim that an entity can successfully defend itself from a DDoS by applying new AV signatures. A DDoS and a malware attack are two completely and separated attack vectors, and they never converge or intersect. Those who reported the described incident, and claimed victory over the cyber attack, do not understand the nature of the attack, and they are rejoicing in a false sense of security. It is perfectly obvious that the report's author(s) do not understand the nature and implications of a DDoS attack, and they completely embrace the fallacy that AV signatures are the panacea for all cyber ailments.

A word of caution it's required at this junction. The mere collection of cyber information on emerging vulnerabilities is not sufficient; the analysis and the design of a mitigation strategy and course of action is what really counts. The mere possession of cyber data does not provide a solution; only the human brain of a true cyber SME can transform that data into an actionable defensive strategy.[41] This is why the main premise of this book states that a cyber conflict is not a battle of cyber systems, but rather a battle of brains. And when it comes to consider the complexity of the cyber spectrum, the mere exercise of brain activity is not sufficient. What it is required is brain activity empowered by cyber knowledge and cyber expertise. There are too many cyber organizations misusing the terms "cyber warrior" and "cyber experts". Cyber expertise is not acquired by virtue of an appointment to a position labeled as "cyber something"; true cyber expertise is earned by individual effort in acquiring cyber knowledge and experience, and it cannot be ascribed as a mere title for a position. If an individual has never confronted a cyber opponent, and survived, that individual is not a cyber warrior. There is a great deal of dignity in being a cyber warrior, and we should not trivialize it.

Let us, then, summarize the premise of this excursus. Every significant piece of cyber code connected to any cyber event, incident or vulnerability, will be disclosed first on open sources, and consequently, can be researched, studied, and analyzed on open sources. To wait for the long process (sometimes as long as four months) of assimilating open sources, as they are being filtered through the different levels of SA feeding into the collective bureaucracy of a government organization, will not achieve any valid purpose. Why? The simply answer is because the cyber data connected to a cyber event, at the technical level, does not need validation or vetting from government entities. Besides, if government organizations lack a certified contingent of cyber SMEs available to examine this cyber data when is unveiled in open source fora, how would the waiting for the process of filtration to bring that data into classified networks would make

---

[41] We can illustrate this point by considering the advertisement of a luxury car company that compares an "ordinary crash test dummy" capable of generating 119 points of data, with their so-called "digital crash test model" capable of generating two million points of data, labeled by this company as their "crash test genius". More data does not create geniuses; only the capability of effectively analyzing and transforming that data into a gaining strategy does.

it any more valid? All that this filtration process accomplishes is to introduce a delay in creating SA regarding the new threat already unveiled in open sources, and by extension, this delay introduces an additional layer of asymmetry by placing us into an even deeper condition of disadvantage.

The answer to this issue is to adopt and nurture the new cyber mindset required to operate in the cyber spectrum. The traditional modus operandi of considering classified information as of greater value than the unclassified counterpart does not apply in cyber.

> Everything that is cyber-related starts at the technical level within a piece of binary code. The information about this binary code appears first in open source.

The correct and prudent process to sustain SA regarding emerging cyber threats is to dedicate cyber SMEs to maintain a vigilant eye on open sources regarding these emerging threats. Through study, research and testing these cyber SMEs can design and implement a defensive strategy and mitigation techniques that are applicable and feasible during the interim period, until a security update is released by the vendors responsible for the affected code.

Cyber asymmetry is not an inevitable condition, but a defeatist choice. Cyber knowledge is freely available to anyone, and those who chose not to learn fall into the abyss of an artificial and self-inflicted asymmetry.

Chapter 7. The ICS case

Within the legal system in the US, and by extension, around the world, there is an abysmal lack of laws codifying what may or may not constitute a cyber conflict. Is a cyber intrusion into the US ICS system any different that a terrorist bomb crippling our critical infrastructure system?

The issue is deeper than a legal issue. The lack of legislation as to what constitute a cyber conflict is based on two primary factors: an endemic lack of cyber knowledge in the judicial field, and a WW2 ideology that conceptualizes war as an aggression based on kinetic and tangible devastating effects. Since the cyber spectrum, in its essence, is neither kinetic nor tangible, and since legislators do not understand the essence of the cyber spectrum, legislators lack the foundation to define what constitutes a cyber conflict, as we already discussed it on the previous chapter 5. The same lack of knowledge exists among the military services, as they attempt to deal with cyber issues, by forcing these issues into traditional warfare doctrines and kinetic warfare templates. I have personally discuss cyber issues with high ranking officers and I was asked on repeated occasions to "translate" cyber issues into a traditional kinetic warfare mold, and even redefine cyber terminology into traditional kinetic warfare jargon. This attitude shows a reluctance to learn a new type of conflict and its corresponding new terminology, thus practically negating the cyber spectrum its distinct and unprecedented entity.

The ICS issue is the perfect case study to grasp the unprecedented nature of cyber conflict. By creating or utilizing a specific binary code designed to exploit an inherent vulnerability on a program written to control and manage critical infrastructure, any individual, independently or in association with a nation-state, will introduce a degree of instability, degradation, or interruption of services in any of the critical infrastructure fields, thus effectively crippling a local, regional or national infrastructure. Nations around the world are presently debating the type of response that a crippling ICS attack deserves. What is the threshold for a national interest response? Do coordinated cyber attacks on networks merit armed forces retaliation?

There is a great disconnect between those who possess the cyber knowledge to engage the issues pertaining to ICS, and those who have the official representation and authority to set laws and policies regarding ICS issues. This statement is based on first hand observations made by this author when attending and presenting at ICS conferences, where various representatives from both government and public sector gather to discuss the cyber threats to the critical infrastructure. Those with authority to make decisions on ICS issues do not have the required cyber knowledge to make educated decisions, and those with the required cyber knowledge are not consulted by those in positions of authority. Following there is a simple set of graphical representations of the current status on cyber ICS issues, illustrating the status of a disconnected ICS community, and the model for an integrated ICS community. There is also a proposed framework solution for the problem at hand.

Disconnected ICS community

The disconnected executive ICS community may integrate many representatives from both the government and the private sector with the majority ownership of the critical infrastructure servicing a national or regional area, thus forming the Industrial Control Systems-Strategic Planning Group (ICS-SPG). However, these representatives do not have the required cyber expert knowledge to address the technical issues affecting ICS. On any typical ICS conference held by the disconnected ICS community there is always an imbalance created by the overemphasis on policies and procedures on regulatory and legislative aspects of ICS issues, to the detriment of the technical aspect of the cyber ICS issues, that are represented during these conferences by cursory reviews presented by individuals with only marginal knowledge on ICS-cyber at best.

The segregated community represented by the Virtual Knowledge Cyber Community (VKC2), where the expert cyber knowledge resides, remains isolated. The VKC2 community not only possesses the cyber knowledge applicable to address and resolved ICS cyber issues, but this community can also transfer an executive-level stratus of knowledge to the ISC-SPG in order to enable this community with the required basic technical knowledge leading to educated decisions on the ICS-cyber complex environment. This transferred knowledge is the Executive ICS Certification (EIC). The integration of these three elements results in the framework of an integrated ICS community, the working model for an ICS enterprise capable of sustaining functionality within acceptable parameters of cyber security. This integrated ICS community operates under the authority of an integrated ICS Advisory Board (I2AB), comprised of executives, ICS SMEs and cyber SMEs.

Integrated ICS community

The importance of having a strategy to protect ICS against cyber threats reaches beyond the protection of the ICS itself, since the functionality of these systems has a direct repercussion on national products, economy, and employment, but even more importantly, on human life, at a very large scale level.

When considering the complex task of defending ICS against cyber threats we need to outline and assess the very unique and critical impact these threats represent. The ICS controls, regulates, and operates very large industrial complexes, critical to national security. Consequently, the assessment on ICS has to reach deeper than simply maintaining their functionality. There are extremely delicate parameters that must remain configured exactly within the original specifications. Any deviation may not affect the availability of a certain product, but it may affect the composition and safety of the same product. The difficulty of accomplishing a deep security assessment of an ICS complex will require the cooperation of professionals and experts in two primary areas; the area in which the ICS under assessment operates, and the area of cyber expertise germane to the ICS environment. This complex interaction represents a difficult challenge, since professionals of dissimilar areas of expertise tend to gravitate toward an attitude of isolation and distrust toward other professionals on a different field.

Another complication is generated by the agenda and primary goal of experts in two dissimilar professional fields. The ones on the ICS arena tend to make functionality their primary task, and by extension they tend to arrange their daily operations around means facilitating the constant availability of the systems under their care and supervision. The arrival of cyber SMEs, with their primary concern and agenda focused on cyber security, may very likely clash with the agenda of their counterparts. A very useful framework to obviate this potential conflict of interest is to use passive penetration testing.

The concept of passive penetration testing (pentest) is to render the assessment process into a non-intrusive, non-destructive operation. Within this framework we can simply assess the presence of cyber weaknesses, without exploiting them. The framework this author proposes is a hybrid ICS assessment (HICSA). Once the passive pentest data is collected, it is fed into a modeling and simulation (mod&sim) exercise, where several scenarios, relevant to the ICS entity under scrutiny, can be conducted.

We can no longer live with the illusion that ICS systems are isolated, and therefore, immune to cyber attacks. Whether or not an ICS system is networked is completely irrelevant, because malicious cyber code can be released either on an isolated network servicing an ICS environment, or on a networked ICS complex reachable through Internet connectivity. The greatest threat in any ICS enterprise is the ubiquitous need for implementing convenient means for those individuals with the responsibility for maintaining ICS functionality. These individuals will most of the time place convenience of use above cyber security, thus creating access points that can be easily exploited. After all, ICS were not designed to operate on interconnected networks, but when we connect an originally designed corporate network hosting an ICS to other networks, or to the Internet, in order to obtain convenient ways to access our particular ICS, at the same time we have created access points for others who are not authorized to connect to an ICS.

The technical information regarding any ICS is readily available on open sources, thus providing the opportunity for any adversary to study the ICS configuration and its potential weaknesses. This is true even in the case of closed ICS environments. Once these weaknesses are discovered, creating a corresponding cyber exploit is within reach of any knowledgeable and determined attacker. And if we add to this equation the factor of implementing access points to the ICS through network connectivity, then we compound the threat exponentially.

Let us examine a brief ICS scenario. Enterprise Alpha operates an ICS environment, and for the sake of the convenience of some administrators, networked access points were created so that such administrators can access the ICS either from their homes, or when they are on the road. This is far more convenient than having to drive all the way from their homes to the Alpha site, in order to control and/or monitor the ICS under their care.

Let us ask a few questions: Did the executives of this enterprise approve these access points, or where they implemented without consultation to the executive body?
If this implementation was approved, was the approval granted after a feasibility study, encompassing a risk assessment? If so, was this feasibility study conducted under the direction of an I2AB? Establishing remote access points to an ICS environment is the second most critical risk. The first is represented by internal threats.

The importance of acting under the direction of the I2AB is to allow the participation of cyber SMEs into the feasibility study, by performing pentests on the ICS protocols germane to the ICS environment under study. It is critical to determine the vulnerabilities associated with the particular protocols enabled in the ICS under assessment. Since all these protocols are inherently insecure[42], there are cyber security measures that need to be incorporated, in order to enable enumeration and monitoring of all active protocols in the ICS environment.

Research conducted by the NSTB program[43] from 2003 through 2009 discovered that ICS environments present themselves as large targets, due to lenient or inexistent network security policies and network configuration errors. These deficiencies are primarily generated by "excessive open ports allowed through firewalls

---

[42]	They are inherently insecure because they were designed within the environment of a close network. Therefore, they lack the security required to operate on an interconnected network allowing remote access. Some of the most common ICS protocols include DNP3, Secure DNP3, EtherCAT, FL-net, Foundation Fieldbus HSE, Modbus, Modbus TCP, OPC AppID, OPC UA, PROFIBUS/ PROFINE.

[43]	"NSTB Assessments Summary Report: Common Industrial Control System Cyber Security Weaknesses", Idaho National Laboratory, May 2010

and unsecured and excessive services listening on them."[44] The routed protocols offered by IPv4[45] are inherently unsecured, and if we add the factor of allowing these protocols to remotely reach into the ICS enterprise, and interact with the unique ICS protocols, even more unsecured that IPv4 protocols, then we have unilaterally managed to lower our defenses and present the ICS enterprise under our responsibility as a highly desirable target to cyber predators.

By extension, and in doing so, we have become an insider threat, and accomplices of the adversary via our negligent behavior. When, and not if, we become victims of a cyber attack against our ICS complex, are we going to compound our dereliction of duty by writing reports with a detailed account of what the cyber attacker did to our AOR, seeking to deflect our guilt by pointing the finger toward the adversary? This has become a very popular escape from our accountability, and today there are literally thousands of reports detailing what the adversary did to intrude into our AOR. When are we going to start writing reports detailing our negligence in failing to protect our AOR?

Connectivity to ICS remote display protocols represents one of the worst weaknesses, because they indiscriminately accept connections from anywhere, and exchange credentials in clear text. When connectivity is allowed from a remote display client, and this client becomes compromised, it becomes a virtual open door to the enterprise ICS display. The attacker now inherits all the types of access granted to the remote client, now no longer under the control of the authorized user, but under the control of a cyber predator. Let us also consider the compound effect of a remote intrusion interacting with ICS applications that have not

---

[44]   Ibid, p. iii

[45]   The routed protocols offered by Internet Protocol version 4 were never designed with cyber security in mind, because they were planned with only functionality in mind, as already discussed in previous chapter 3. Among them we have the legacy unsecured protocols such as FTP, Telnet, HTTP, that can easily be replace with the corresponding secure versions.

been patched because the ICS complex was normally operational as a closed network. However, after an intrusion has created an open door into the ICS enterprise via a remote connection, the unpatched weaknesses become an enhanced set of exposed vulnerabilities.

Since remote connectivity becomes a much desired convenience that very seldom will be denied, it follows that the proper authorization to implement remote connectivity should be granted only when two conditions are met. The connection between the ICS local enterprise network and the remote client must establish connectivity via a VPN[46] authorized secure channel, and cyber security monitoring software must be installed on both endpoints. The monitoring of all VPN connections must implement strong authentication procedures, and certified cyber SMEs must be employed to perform the monitoring of the raw network data exchange between the endpoints.

Once we establish the monitoring of all inbound and outbound traffic to and from an ICS environment, then it's time to establish a base line for all authorized traffic, in order to identify anomalous traffic, that may or may not be authorized. The best possible ways to establish this base line is by identifying all protocols required for the functionality of the iCS environment, and disable all others, through the establishing and implementation of access policies.

Many organizations do have their access policies, but they remain as written statements lacking enforcement. But the enforcement of written access policies on an ICS requires the deployment of authentication and authorization procedures. The goal is to verify a user's identity, and provide access level based primarily on a clearly defined set of user's role, and the corresponding privilege level required for performing the duties associated with that role. Furthermore, we must keep logs of all access events in order to audit any infraction, intended or accidental. ICS devices such as RTUs and PLCs operate on a diametrically opposite posture to these outlined fundamental security principles, and all the common

---

[46]   Virtual Private Network

ICS protocols lack strong authentication procedures. Very often ICS are void of the required capabilities to implement a secure and granular access control designed to match the degree of access corresponding to the role of a particular individual. Users are given unrestricted access regardless of their official roles. And to add insult to injury, the majority of ICS devices do not log events in order to save the limited memory available to them.

Let's see a real case showing the results of operating on a disconnected ICS community, without an I2AB. On January 2012, a web site[47] offering news and commentaries on IT management in the federal government, reported a cyber incident affecting a Northwest railway computer network. The cyber attack was characterized as originating overseas, and its effects as lasting two days. An original memo issued by a government agency stated that the cyber attack was considered to have an origin overseas, but a spokesperson for a railroad association expressed disagreement by saying that the memo characterization of the origin of the cyber attack was incorrect, and the cyber incident was not a targeted attack, and the same criticism was voiced on another online report.[48] Was there any cyber security team monitoring the network traffic corresponding to the affected railroad entity? Was there any report issued by this cyber security team regarding the anomalous network traffic responsible for the disruption?

What this case illustrates, in addition to the irresponsible and negligent characterization of cyber incidents, is the discernible pattern that begins to emerge when we consider similar other cases in a collective analysis. The emerging pattern highlights the pervasive escapism exhibited by the responsible parties, deflecting their accountability for negligent behavior in failing to secure their ICS environment, and thrusting into the public spotlight an imaginary foreign adversary. These negligent and

---

47   http://www.nextgov.com/nextgov/ng_20120123_3941. php?oref=topstory.
48   http://news.cnet.com/8301-1009_3-57366341-83/dhs-disputes-memo-on-purpoted-railway-computer-breach/

ignorant spokespersons customarily issue written statement to the press with hyperbolic accounts of cyber attacks perpetrated by imaginary nemeses from overseas.

Just a month before the Norwest railroad case, on November 2011, another cyber incident was described by a spokesperson, an alleged prominent expert on ICS protection, who characterized the event as a foreign attack on a water facility in Springfield, Illinois. The cyber attack was described as perpetuated with the aid of stolen credentials, and the spokesperson even anticipated that other ICS environments may become the target of such cyber attacks by the same group of Russian perpetrators. Let's fast forward to December 2011, when we discover that the statement of the alleged prominent expert was based on an unsubstantiated hypothesis built in his own mind, and released to the press before gathering and examining the facts associated with the "cyber incident." The cyber activity associated with this event was a legitimate action, conducted by a legitimate contractor who was instructed to conduct a remote checking on the SCADA system installed at the water facility, while he was vacationing in Russia.[49]

When analyzing a cyber event the qualified SMEs rely solely on cyber forensic facts, and then they present a characterization that corresponds to the collected facts. Beyond these facts, there is very little room for personal opinions and groundless hypotheses in cyber analysis. So, let us ask ourselves a few questions regarding the previously cited case of the railroad cyber event : what are the cyber credentials of the originators of the two conflicting opinions? Were these opinions supported by cyber forensic evidence? On these two opposing opinions only one can be correct, and that is the one supported by the network traffic logs. Were these two individuals capable of reading and interpreting network traffic logs? Did the victim organization have any of these required logs? Were those who examined the IP addresses associated with the cyber incident qualified to properly

---

[49]   http://www.wired.com/threatlevel/2011/11/water-pump-hack-mystery-solved/

resolve them? Resolving IP addresses is a technical skill that requires knowledge of networking technologies, and those who lack this knowledge frequently incurred in erroneous interpretation of IP resolution techniques.

According to the web site report, there was a significant number of representatives from technology organizations and government agencies meeting to discuss the cyber incident. What was the overall cost incurred by the attendance of all these representatives? Did this meeting convene before or after the discovery that the cyber event was not a targeted attack? Were the alerts sent to several hundreds railroad companies and transportation agencies disseminated before or after discovering the inaccuracy of the government memo? All this questions are extremely relevant when considering the amount of time and money invested in arriving at the final conclusion, and all the apparent confusion surrounding the investigation of the cyber incident. The presence, participation and guidance of an I2AB was definitely lacking in this case, as it is obvious by the overall confusion and lack of specificity surrounding the whole event.

This railroad cyber incident is a serious event, illustrating the many aspects of the complexity involved in ICS matters, and the paramount importance of supporting any ICS enterprise with a qualified VKC2, and the guidance and endorsement of final cyber analyses provided by qualified SMEs, working in synergy within an I2AB body. ICS is very critical, and it must be in the capable hands of qualified professionals, not in the hands of neophytes. There are too many examples of inept handling on cyber issues, and the price tag paid for ineptitude is on the rise. Even more important, the self-induced asymmetry level continues to escalate, simply because the ICS vulnerabilities are numerous, and many of those within the ICS environment in general maintain an indolent attitude toward the impact of cyber attacks targeting their particular ICS area. Who is being held accountable for the impact of this type of cyber events on human life? Cyber events affecting the ICS complex always have a high impact on human life, on economy, and on national security.

There are primarily two schools of thoughts responsible in part for the apathy regarding the cyber impact on ICS environments. One that maintains a veil in front of their eyes, assuming that the original design of ICS as an isolated system still remains isolated from the Internet. The other, assuming that connectivity to the Internet is privately controlled and only partially connected. This last statement is either a smoke screen, to obscure the reality of the issue, or a statement based on networking ignorance. Internet connectivity is an absolute; either you are connected, or you are not. There is no gray area in between. Once a system is connected, it is virtually reachable from anywhere. The only degree of isolation is achieved via specific cyber software and hardware, regulating access control policies that have to be carefully designed, and monitored, by qualified cyber SMEs. Anything outside and beyond these parameters is a wishful and gullible fantasy.

An interesting online article[50] reports that a student researcher established the location of more than 10,000 ICS entities connected to the Internet, exhibiting various degrees of vulnerability to cyber attacks, as a result of their lenient cyber security practices. The critical finding was the fact that the researcher determined that only 17 percent of the ICS systems he discovered connected to the Internet required authentication before authorizing connection. What about the remaining 83 percent? The administrators of such systems were either unaware they were connected to the Internet, or they were negligent of installing access control systems. The primary tool[51] used by this researcher is freely available online, and the corollary is this: if a student can in a few months gather such important intelligence data, exposing the cyber vulnerabilities of great number of ICS environments, what can a nation-state accomplish, in preparation to a massive cyber attack?

---

[50] "10K Reasons to Worry About Critical Infrastructure", Kim Zetter, January 24, 2012, http://www.wired.com/threatlevel/2012/01/10000-control-systems-online

[51] The SHODAN search engine, developed by John Matherly

On two occasions, at the end of 2010 and 2011, the ICS Cyber Emergency Response Team (ICS-CERT) released urgent alerts[52] regarding the increase use of SHODAN to locate vulnerable ICS sites, with numerous cases where remote access has been allowed via Internet connectivity void of access control devices, and access configuration with default or weak user names and passwords, thus enabling cyber intruders with unrestricted access to their ICS facilities, since default account names and passwords are publicly available.

The September 2011 report of thousands of vulnerable ICS devices discovered with the aid of SHODAN is neither the first, nor will it be the last. A previous report in February 2011 highlighted the same usage of default names and passwords.[53] One may ask: how many of the ICS devices discovered in the February report appeared again in the September report? How many ICS enterprises acknowledged the 2010 and 2011 ICS-CERT alerts, and implemented the necessary corrective measures?

Those who remain apathetic to all ICS cyber alerts will wake up from their slumber by the collapsing sound of their compromised cyber infrastructure when the adversaries find the access door, wide open, into their ICS kingdom. For the rest of us, who heed these cyber alerts and remain vigilant, the presence and proliferation of cyber tools should not lead us into a state of panic and desperation. Why? Because cyber tools are intrinsically neutral, and their use is subject to the ethical orientation and agenda of the user. Let's reiterate once again the premise of this book: cyber is an equalizing force. The same cyber tool that an adversary can use against our network enterprise, we can use it to defend and protect it from that adversary. Thus, the winning factor resides in the brain of the human being operating the tool. The more that person knows about cyber in general, and the

---

[52] ICS-ALERT-11-343-01—CONTROL SYSTEM INTERNET ACCESSIBILITY, Dec 09, 2011
[53] Two other reports, in April and November 2011, highlighted the same problem. See ICS-ALERT-11-343-01

more that person knows about the particular tool being used, the greater the advantage over the opponent.

On 2009 a study on the effect of a cascading effect attack on the Western US power grid was published by a couple of Chinese students.[54] Their research is based on a model of multiple interconnected power nodes on the western US electrical power grid, illustrated in the simplified representation below:

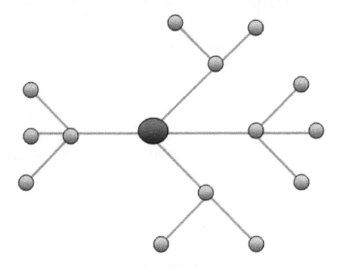

The traditional view is that an attack on the node sustaining the highest load (represented by the bigger node) is the most likely to trigger a cascading effect of consecutive overload failures, eventually leading to the collapsing of the power grid. The findings of the Chinese researchers, however, contradict this traditional view, and shows that the greater impact of a cascading effect are very likely triggered by a failure on the nodes with lower loads. Thus, an attack on the nodes with lower loads will yield the greater results in collapsing the western US electrical power grid.

---

[54] "Cascade-based attack vulnerability on the US power grid", Jian-Wei Wang, Li-Li Rong, Institute of System Engineering, Dalian University of Technology, PR China, 2009. The study was supported by the National Natural Science Foundation of China under Grant Nos. 70571011 and 70771016.

With this information at hand there is a great potential for reviewing this study with the aid of mod&sim technology.[55] If the results are confirmed, then we have a great opportunity for applying anticipatory analysis, and prepare a sensible remediation plan to avoid the disastrous results of a cascading effect affecting our nation's power grid. The main concept to maintain control, safety, and availability of our ICS entities is to take the initiative on strategic planning, when confronted with credible and technically feasible cyber vulnerabilities. At the present time, there is an abundance of demagogy and official reports that do not translate into strategic changes in the way we deal with the ICS threat. Cyber incidents are preventable, but we are not preventing them. This is a dangerous status quo nurtured by indolence, and the painful consequences of major regional ICS failures are quickly forgotten. However, the impact of ICS failures is usually measured in human tragedies and human life, and in ecological and economical disasters. We have the power to mitigate ICS vulnerabilities, but are we willing to do it?

---

[55] See chapter 16 ahead

Chapter 8. The Certification and Accreditation myth

This is a brief chapter dedicated to highlight a very simple but important fact. The Certification & Accreditation (C&A) process is a positive initiative designed to establish a formal base line for the cyber security of an information system. The candidate system must demonstrate compliance with this base line in order to receive an official authorization to operate in an approved environment, under approved security policy parameters. However, while a system who earns a C&A approval is in compliance with the security policies of the host environment, it is not immune to cyber attacks once it becomes operational. This distinction is very important, but very often obfuscated or ignored.

C&A, when considered in isolation, is to cyber defense what the Maginot Line was to the French defense; a voluminous, expensive, and often inefficient process that do not always contribute to a line of cyber defense. Those who put their trust in C&A as if it was an unassailable armor fail to understand that cyber conflict is a very dynamic conflict, not a static one. In the C&A process, a static and isolated cyber device is measured against a cyber security baseline, designed to ensure that a cyber device meets the basic parameters outlined by the security policy of an organization.

Thus, the C&A process is intended to ensure that cyber systems comply with formal and established security requirements, properly documented and authorized, and endorsed by a particular organization. The intent of the C&A process is to ensure that all cyber systems are deployed only after undergoing the scrutiny and approval declaring them as secure cyber systems.

However, the C&A scrutiny consists most of the time of a paper audit completely detached from a technical audit, and the whole process is reduced to a declaration that the C&A candidate meets the adopted and authorized security requirements. Regretfully, nothing in the C&A process guarantees that the C&A candidate system will maintain the accredited security posture throughout its life cycle. Quite often the C&A process is taken simply as a mandatory paper work requirement needed to initially deploy a cyber system, and forgotten once the process is complete. This is the fallacy of the C&A, failing to adhere to its intended original design, as a valuable tool to manage the security of a system throughout its life cycle.

The C&A process is fundamentally a managerial process that centers on documentation[56], not on technical analysis of the candidate system. There are long series of checklists and tables whose purpose is to provide a quantifiable measurement of confidence that the candidate system will operate in accordance to a specific cyber security policy. The certification factor seeks to validate through comprehensive documentation that the candidate system will comply with the security policy mandated in the host environment, while the accreditation factor evaluates the risk assessment caveats outlined during the certification process, and either accepts or rejects such assessed risks. The accepted guidelines for C&A do not call for an exhaustive and technical scrutiny of the candidate system, but rather a cursory review of the potential threats perceived as relevant to the system under scrutiny. Vetted guidelines mention that the Certification Authority (CA) should exhibit a basic familiarity of the relevant threats to the system under scrutiny, and in the majority of the cases, consultation of generic threat information will suffice. If any pen tests are administered during the C&A process, they are done under ideally controlled circumstances, which become irrelevant after the system is connected to the local enterprise network, and by extension, to the Internet.

---

[56] During the certification process the accepted guidelines call for the consultation of baseline documentation addressing security issues, including security policies and system architecture.

Here is where the C&A myth comes into play. After a candidate system has received the letter of accreditation and either the Authorization to Operate (ATO) or the Interim Approval to Operate (IATO), the approve cyber system is connected to the Local Area Network (LAN), Wide Area Network (WAN) and to the Internet, this very same system becomes a different entity than the one evaluated under internal and ideal circumstances, as an isolated system. The access controls and the port configuration applied to the new system will transform the behavior of this newly connected system, by becoming a dynamic entity once its connectivity is enabled. Many may argue that the C&A process also includes the factor of accreditation maintenance, intended to ensure that the accredited system continues to operate in compliance with the security parameters outlined in the accreditation letter. This is a stipulation that is seldom observed.

Only U.S. federal departments and agencies mandate C&A, and private organizations are not included in this mandatory C&A process. However, about 90 percent of the nation's critical infrastructure resides on private networks that are not part of any U.S. federal department or agency, and the cyber systems running critical infrastructure components (electrical power, chemical, nuclear, transportation, telecommunication, banking and finances, agricultural, food and water supply entities) are not mandated to adhere to the C&A process.

If there are no compulsory regulations for C&A on the ICS private organizations, then there are no cyber security policies providing a baseline for security on the systems controlling the critical infrastructure. But for those organizations that operate under mandatory compliance with the C&A process, there is a false sense of security by assuming that the accredited systems are invulnerable. The corollary is this: once a system receives ATO status and is connected to a network, it becomes a new dynamic entity, exposed to all the dangers and threats hovering on the Internet. The base line provided by the C&A process is only the starting point, a lower basic level of cyber security. Many other layers of cyber security are necessary, and must be continually implemented, in order to continue protecting the system through its lifecycle. To ignore or discard this reality is an invitation for disaster.

Chapter 9. The best of both worlds

The detection and neutralization of cyber attacks is always a difficult task, but it becomes even more difficult when this task is surrounded by the numerous misconceptions imposed by neophytes transplanted into the realm of cyber security. Driven by the intimidation of what they don't understand, they sacrifice knowledge on the altar of ignorance, and they blindly entrust their aspirations for cyber security to automated devices. The overarching premise in cyber defense is the factual realization of the magnitude of the task of defending an enclave against a myriad of exploits launched against it. Those who remain in the periphery of the cognitive envelope of cyber understanding are easily overwhelmed by the magnitude and frequency of cyber attacks, and find solace in dreaming of powerful automated detection systems that will protect them from the attacks they do not understand. Those of us who work inside the cognitive envelope of cyber understanding do not despair at the sight of the avalanche of cyber exploits. Instead, we contribute to the design and building of tailored cyber defenses, based on a layered cyber security posture that integrates both automated detection systems, working in synergy with cyber security strategies designed by certified and experienced cyber warriors, and deployed and implemented under the direction of cyber SMEs. The best of both worlds is the paradigm of fusing the work of automated detection system and the analytical skills of cyber SMEs, in a unified task where the human being, not the automated system, has the last word in cyber defense.

This author recently spoke at a meeting of cyber representatives from a selection of several government and industry organizations, and the perennial question of the best COA was brought for discussion. The adoption of the best combination of automated IDS and IPS, in conjunction with education and training for users, preparing them on how to avoid falling victims of cyber predators, quickly became the consensus and focus of attention. After confirming that this was the preferred solution by the audience, this author requested the opportunity to present a counterargument, in order to demonstrate the fallacy of this proposed solution, born of the endemic misconception about the nature of a cyber confrontation.

A cyber conflict is a battle of brains, not of machines

The counterargument is based on the inherent instability of the preferred solution, labeled by this author as "the two-legged table". This paradigm, as conceptualized by the audience, and established on exclusive reliance on the two components formed by automated IDS-IPS and the training program, is as unstable as a two-legged table, because it represents an incomplete platform. The minimum requirement for stability requires three components, and the missing component in this faulty paradigm is the presence of the certified cyber SMEs, qualified and experienced in cyber conflicts. It is extremely unsettling to be in the presence of individuals who honestly consider the two-legged approach the answer to their cyber security needs. How did they all begin to exclude the cyber SME from their cyber security equation? How do they begin to conceptualize that the answer to a cyber incident resides in piling more and more automated cyber defense equipment in front of the adversary, while training users on safe cyber guidelines? This is not better than the Maginot solution.

The answer to a stable, sustainable and resilient cyber security paradigm is based on a three-pronged approach, hinged on the presence and direction of qualified cyber SME with experience in cyber conflict operations. They are the brains in the cyber conflict, and the automated cyber defense systems are simply their tools. On the other side of the conflict there is a human opponent, designing and coordinating offensive strategies, and only another human defender can detect and analyze the offensive cyber strategy, and design countermeasures to protect the targeted network. When one thinks about the possible answer to the widespread preference of so many people favoring the faulty two-legged approach, the possible answers are discouraging: is it because we still don't understand the essence of a cyber conflict, or is it because we lack in our enterprise the availability of cyber SMEs qualified for cyber conflict operations, and therefore we put our trust on automated equipment because the vendors of such equipment promise us they are sufficient to defend our AOR? Are we that gullible? Are we that desperate? What are we planning to do next, when confronted by a determined and skilled adversary; throw pixie dust, and hope for our vulnerable network to fix itself?

Let us restore a solid foundation for our cyber security paradigm. It must be based on qualified cyber security professionals, participating in the design of cyber security policies, and in charge of implementing, maintaining and operating the cyber security devices deployed to protect our AOR. The daily cyber security operations must be under the control of qualified cyber SMEs with knowledge and experience in cyber conflict operations. In this paradigm there is place for a productive and synergistic three-prong approach, including the cyber SMEs, with operational supervision over the cyber automated devices, and with a dynamic and relevant cyber security training program for users.

The term "cyber warrior" has been grossly trivialized by many organizations, ascribing this term to any individual assimilated by metamorphosis,[57] whose working duties are broadly defined as cyber activities, even if they remain in the periphery of the cyber envelope. The term "κυβερ πολεμιστής"[58] (cyber warrior) should be reserved only for those who operate within the cyber

---

[57] See chapter 4

[58] This term, coined by this author, transliterates into "cyber polemistis", and should be ascribed only to those who have an in-depth understanding of cyber security and put their knowledge and experience to the service of a higher cause to implement and improve cyber security. The "κυβερ πολεμιστής" shield is also an adaptation by this author.

spectrum at a professional level of cyber security understanding, with empirical knowledge on cyber conflicts, operating within parameters defined by cyber expertise, ingenuity and creativity in designing ad hoc defenses against ad hoc cyber threats.

Those who simply work in front of a monitoring console, in a passive role of observing and reporting what an autonomous and automated IDS or IPS does, do not qualified for the appellative of cyber warrior. This title is only reserved for those who are engaged in an active and proactive cyber defense, for those who actually encounter and counteract a cyber adversary's strategies with a counter strategy, designed to deny or degrade the initiatives of the adversary, as the cyber conflict unravels. These counter strategies are designed and implemented by the cyber warrior, not the result of a canned solution indiscriminately applied by an automated cyber security system.

The most devastating and inherent cyber weakness is generated from a general attitude of complacency and triviality, permeating every stratus of any given organization, and tolerated across all levels of leadership. The general view is to consider cyber spectrum as a virtual arena designed for our convenience and entertainment. This school of thought trivializes every aspect of cyber activity, and introduces the greatest risk factor in our combined and interconnected cyber existences. This mindset precludes the creation, sustainment, and deployment of a contingent of true cyber warriors.

A contingent of "κυβερ πολεμιστής" (KP) cyber warriors can only emerge from a mindset that requires a radical and sacrificial metamorphosis. The KP culture can only be formed by individuals that truly understand the lethality of the cyber spectrum, and commit themselves to a career of discipline and heightened sense of duty, renouncing to the trivial and complacent attitude sponsored and cherished by the masses that traverse cyber space as if it was a virtual entertainment park.

The cyber spectrum is an extremely contested environment, and it's very unwise to venture into this realm without the proper defense strategy. Because of the pervasive lack of knowledge regarding the benefits and dangers of traversing the cyber spectrum, many institutions resort to depend exclusively on automated intrusion detection systems, which are offered as the panacea against all cyber ailments. This strategy is self-defeating and myopic, since depends exclusively on a device that is both flat and shallow in its approach to intrusion detection and cyber defense. Automated network intrusion detection systems (NIDS) do offer a tremendous advantage in terms of speed for processing network traffic data, but speed in itself will not bring victory. Human assets working in conjunction with automated intrusion detection system offer the depth and multi-dimensional analytical perspective lacking on NIDS. Cyber security and cyber defense should not be subject to the fallacy of an "either-or" approach; the winning strategy resides on the "both-and" approach of using both the NIDS and the human cyber security experts.

A recent paradigm in the IT industry has been the design and deployment of supplementary defense architectures knows as

Intrusion Prevention Systems (IPS), in addition to the traditional IDS systems. The theory is to enhance a cyber defense posture by deploying both IDS and IPS. This concept springs from the concept that IDS are reactive in nature, while IPS are allegedly proactive, and therefore portrayed as capable of mounting an active defense. Since the general concept behind IDS is a set of signatures matching known threat strings, the IPS solution is offered as an improvement, operating on the foundation that an IPS will detect anomalous network traffic heuristically identified as containing new threats, even before their threat signatures become a known entity. This approach has some limited value and success, but IPS vendors usually hyperbolized their claims for cyber protection.

The success of IPS is overrated at best. These systems have a valid place in the cyber security equation, but they are far from representing the proverbial "silver bullet" that kills all cyber threats. They do not constitute the panacea for all cyber ailments. The best of both worlds is constituted by the equation where we integrate into our layered cyber security posture both the automated mechanisms (IDS, IPS) and the superb element of the human mind operating within the creative and proactive mission parameters of a seasoned, qualified, experienced and dedicated cyber warrior KP type. Thus the equation for cyber defense in depth can be defined as:

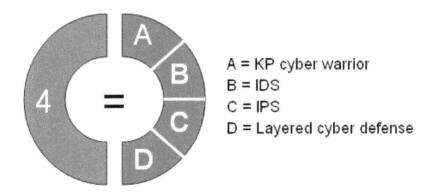

A = KP cyber warrior
B = IDS
C = IPS
D = Layered cyber defense

4 = Cyber Defense in Depth

This paradigm clearly restores the artificially displaced role of the KP assets as the main element in the equation. This displacement is the sub product of the erroneous dependence on automated systems thought as the answer to the cyber defense dilemma by those who consider an automated system as superior than the human mind. This misleading concept results when those outside the cognitive sphere of cyber defense mistake speed and volume of processing with superior discerning power. Automated systems can only process preconceived and preprogrammed instructions, and they are unable to correlate and design new courses of action (COA) when confronted with unforeseen or unknown circumstances. Only the human mind is capable of establishing correlations and design new COA germane to new threats.

Chapter 10. Cyber superiority

Achieving cyber superiority is an ephemeral goal, since true cyber superiority is unattainable. Only an acceptable degree of cyber superiority can be achieved, and this, only as a condition on a given point in time, not as a historical continuum. The cyber spectrum, the result of all cyber factors at a given time, cannot be controlled and mastered as a condition, but only as an event. The cyber factors leading to a cyber threat event are too numerous and globally distributed to remain under the control of any given nation.

Cyber superiority should be understood as a process, not as a goal. Whoever achieves the highest degree of control over the cyber factors characterizing a given historical cyber event, that entity can be considered as having cyber superiority on that cyber event in particular. Maintaining such condition of cyber superiority on any given future cyber event will require maintaining a complex process of preparation and research germane to a virtually unpredictable series of future cyber events.

A cyber event may originate anywhere around the globe, and because of the unpredictability of the factors combined on any given cyber threat, the process of cyber superiority requires extensive knowledge of all potential cyber exploits capable of harming one's enclave. Thus, knowledge of the characteristics and weaknesses of one's enclave is imperative. The possibility of achieving an acceptable degree of cyber superiority can only exist as a layer built on the foundation of a sustainable cyber defense posture. Therefore, a sustainable cyber defense posture is a condition for cyber superiority. A general knowledge of the cyber spectrum is insufficient to aspire for cyber superiority. A specific knowledge of the cyber characteristics and weaknesses on one's enclave is the only acceptable foundation for aspiring to cyber superiority.

The concept of cyber superiority hinges on the correct understanding of the unique essence of the cyber spectrum. Attempts to define cyber superiority must begin with a concise and exact definition of cyber, and since we have already exposed the pandemic lack of knowledge about the true essence of cyber, we are surrounded by vague and unsubstantial statements about cyber superiority, the product of unqualified opinions by spokespersons lacking understanding on the cyber entity.

Let us take a random sample of such unqualified opinions available online. One such misguided opinion[59] offers a plan to provide a government entity with a model of cyber superiority by applying a particular dynamic radio frequency resource management for EMS exploitation in order to prevent cyber attacks, and ensure unrestricted operational capabilities in the cyber arena. Let us review the cyber principles outlined in this book to elucidate the degree of irrelevancy present in this online opinion.

EMS is the continuum of all electromagnetic waves arranged according to frequency and wavelength. EMS is neither included in the cyber spectrum, nor the cyber spectrum included in EMS; the former is essentially an array of radiation waves, while the latter is essentially a collection of binary codes. EMS can and does co-exist with the cyber spectrum, but they are and remain two distinct spectra.

EMS remains as the domain of waves, and cyber remain as the domain of cyber code.

The essence of the difference between EMS and cyber spectrum is the primary unit of information. EMS generates radio waves, while cyber spectrum generates cyber code.

The quoted misguided opinion incurs, in a very short paragraph, in three fundamental errors regarding cyber superiority. First, it

[59]   http://www.androcs.com/html/cyber_security.html,   2010   ANDRO Computational Solutions, LLC

fails to distinguish the essential differences between EMS and cyber. Second, it erroneously equates cyber superiority with cyber defense, and third, it erroneously equates cyber superiority with cyber availability. Cyber availability is a fundamental condition of existence in the cyber spectrum, and cyber defense is a fundamental condition for cyber sustainability. Cyber superiority is an operational condition rooted on the pre-requirements of cyber availability and cyber sustainability, but cyber superiority transcends these two elemental conditions.

Cyber superiority is the result of a dynamic and complex interaction of several factors resting on the foundation of cyber availability and cyber sustainability. Cyber superiority requires cyber intelligence (CYBINT), human intelligence (HUMINT), and signal intelligence (SIGINT), in addition to a complex network of global (and sometimes risky) access agreements. Cyber superiority requires a delicate balance of strategic and tactical measures that work better when under the supervision and operational control of decentralized, but interconnected, fusion cells formed by cyber SMEs. Centralized hierarchies, with layers of leadership void of technical cyber knowledge, often become a layer of latency that is counterproductive to operational success. Cyber superiority can only be measured in the context of a cyber conflict. Cyber superiority does not exist in the vacuum of cyber inactivity, where it remains only as a probability, not a certainty. Therefore, the requirement for CYBINT is a must in the quest for cyber superiority. However, we cannot even begin to aspire for cyber superiority, if we cannot discern and understand the essence of the cyber spectrum. The entities that are seeking for guidance on this quest should focus their attention on qualified cyber SMEs, and avoid those who are incapable of discerning the very essence of a cyber conflict, and those who offer misguided advice from behind the blinding veil of cyber ignorance.

Cyber superiority requires the foundation of cyber technical knowledge at a level superior to that of all potential cyber adversaries, but it transcends pure technical knowledge. It requires knowledge of the adversary capabilities, specific knowledge of the opponent's cyber infrastructure, and an inventory of the cyber

SMEs supporting the rival side. The quest for cyber superiority begins with the laying of the foundation of CYBINT, a new and unprecedented form of intelligence gathering that requires the cooperation of HUMINT and SIGINT. Only qualified and experienced cyber SMEs are capable of discerning the potential value available in collected data associated with cyber activity and cyber operations conducted by potential adversaries. The data from HUMINT and SIGINT may carry valuable information that only a qualified cyber SMEs will correlate, via an analytical process germane only to those who can harvest the kind of information containing strategic and tactical value in the cyber arena.

Cyber superiority requires instinct, anticipatory analysis, trend analysis, even deviousness, because a cyber conflict is never a scripted exchange, where team A follows a previously determined course of attack, and team D follows an expected course of defense. Regretfully, many cyber exercises follow this scripted model, simply because the opponents are not prepared to conduct a realistic exercise, where creativity, unpredictability, anticipation and deviousness are present. This author participated in one such exercise, with team A representing the cyber attackers, and team D represent the cyber defenders, with this author as the team lead. Team A's plan was based on the assumption that, by depriving team D of the automated alerts displayed on a console receiving a feed of data from an IDS, it would result in team A achieving a successful cyber attack by depriving team D of a baseline of defense. What team A was not expecting, because they were counting on a passive defense depending exclusively on the display of automated alerts, was the defensive strategy used by team D, based on SQL queries polling directly from the IDS database, bypassing the IDS console, thus extracting the data representing the attack strategy of team A. Once the SQL queries revealed to team D the attack mounted by team A, it was trivial for team D to nullify the attack from team A. At the end of the exercise, when an analysis of the actions of both team was examined, team A claimed that the victory of team D should be disqualified, because team D used a defensive strategy that was not part of the scripted exercise. The final result was to collect the strategy used by team D and incorporate it on upcoming exercises, but this author was excluded from participating in future exercises.

A cyber conflict is never a scripted exercise, because there are always human brains behind the keyboards, and whomever has the best command of cyber knowledge and of the cyber devices involved in the conflict, that opponent will achieve victory. Cyber instinct, based on knowledge and experience, and aided by CYBINT, is a powerful recipe for the probability of victory and cyber superiority. There are numerous factors leading to achieving the condition of cyber superiority, though only on a given point in time. Some of the elements contributing to the overall aggregate of cyber superiority include cyber knowledge (theoretical and empirical), expertise, creativity, ingenuity, instinct, initiative, and some relevant aspects of critical thinking[60] (pattern analysis, trend analysis, anticipatory analysis, semiotic analysis). Once these elements are fused into a cohesive synthesis, cyber superiority emerges as a condition that transcends the sum of the fused elements.

Nowhere is the concept of nonlinearity more important than in analyzing a cyber conflict, where the whole is always greater than the sum of the known elements comprising the particular cyber conflict at hand. The complexity of the cyber operating environment requires a nonlinear analysis, because the different elements always appear is a state of chaotic flux. However, a seasoned cyber analyst can discern the nonlinear order present in the dynamic of the cyber conflict. Complex adaptive systems are omnipresent during a cyber conflict, but they are also evident, and not obfuscated. The manifestation of achieving a degree of cyber superiority is the capacity to discern those elements in the midst of the apparent chaotic flux.

Performing under the guidelines of critical thinking principles is not jus an advantage, but rather an imperative in cyber conflict,

---

[60] Properly applied, after tailoring them into the cyber arena, from the seminal concepts articulated in Intelligence Analysis, Wayne M. Hall and Gary Citrenbaum, ABC-CLIO, 2010

if we are to overcome the reactionary vicious cycle in which we are currently engaged in our customary cyber defense models. Capitalizing on the principles of critical thinking will lead to anticipate the actions of the cyber opponent, and even nullify the advantage that such adversary might have gained during the early stages of the cyber attack. The advantage in cyber analysis is the availability of cyber forensic evidence, but the downfall is the scarcity of qualified cyber SMEs capable of discerning all the observables present in the cyber evidence. One of the paramount tenets of critical thinking in the cyber arena is the premise of finding undeniable facts, and once analyzed, use that evidence to deploy countermeasures, thus regaining the initiative. No conflict in history has ever been won by simply reacting to the initiatives of the adversary. The only way to victory is by analyzing the facts and the strategy of the adversary, taking actions conducive to take that initiative away from the opponent, and impose our initiative on him. Why then do we hold to the fallacy that in cyber is any different? Simply reacting to the actions of the adversary will never propel us into acquiring cyber superiority, and cyber victory will always be a perennial mirage.

The proper and accurate analysis of cyber evidence comes with a price; time and resources invested in learning how to properly conduct forensic analysis of the cyber data at hand, and additionally, the knowledge of where to find such forensic evidence. The individual that cried wolf in the case of the Illinois water facility, by stating there were Russian perpetrators behind the cyber attack targeting the water facility, did he know there was forensic data available? Did he attempt to contact the cyber SMEs who were capable of locating the forensic data, and read such data? After creating an alarmist situation, and disseminating that alarm to other water facilities, warning them of the impending Russian attack, the undeniable facts surfaced showing that there was no such cyber attack.

And in the case of an actual cyber attack, if we don't know how to identify the evidence characterizing the prelude of such threat, we will be condemned to always take a reactive posture, always dancing at the tune imposed by our cyber adversary, and reducing all our efforts to write and disseminate post-mortem reports. This author has used anticipatory analysis to identify imminent cyber attacks, and in conjunction with anomaly analysis, has been able to create proactive cyber strategies to anticipate the actions of the opponent, and regain the initiative by implementing a successful proactive cyber defense model.

In a cyber conflict there are only two options; victory or defeat. Cyber, as the great equalizer, provides both opponents with the same opportunities and the same tools. The defining factor is not on the cyber devices, but on the human assets behind the keyboard. The legitimate cyber warrior that invests the time and effort in seeking cyber superiority by paying the required price has a better possibility of achieving superiority and victory. It all depends on commitment and determination to be the best. The military and economical prowess of a nation is nullified if it lacks cyber brain power.

Chapter 11. Anatomy of a cyber intrusion

Let's have a quick show of hands among the readership of this book, to ascertain the level of preparation that any given reader may have to offer when confronted with the unavoidable fact of a cyber intrusion. I ask every reader to mark with a circle the number of ports they recognize in the list below:

- ICMP (portless)
- 0
- 1-20
- 20
- 21
- 22
- 25
- 50, 500
- 53
- 67 udp
- 68 udp
- 80, 443

- 111
- 161
- 162
- 179
- 445
- 502
- 1433, 1434
- 1521
- 3128
- 3389
- 4500
- 6667-7000

Once the reader has finished circling the recognized ports, please place a check mark on those ports the reader considers to be particularly associated with potentially malicious traffic, when meeting other suspicious circumstances. The results of this impromptu statistics will illustrate the core issues germane to the analysis and remediation of a cyber intrusion.

A cyber defense program based on a reactive paradigm is not a cyber defense program at all. The only acceptable cyber defense program must be based on a proactive and preventive paradigm.

A reactive cyber defense program is both a contradiction of terms, and a waste of time and resources. Purely reacting to cyber intrusions is equivalent to the exercise of a dog chasing its own tale; it's futile and counterproductive.

If an enterprise doesn't know how to assess its own cyber weaknesses, any defensive attempt is pointless. Cyber operations are based on cyber knowledge, and cyber conflicts are won based on dominant cyber knowledge and cyber expertise, because any cyber conflict is really a battle of brains. One of the most devastating and common practices in the cyber spectrum is the implementation and deployment of cyber enterprises that connect their networks to the Internet with default configurations applied to their core network devices. This modus operandi is quite common when cyber functionality is place above cyber security. Let us review a few best practices that are consistently ignored when sacrificing cyber security in the altar of functionality.

> The only acceptable cyber defense program must be based on a proactive and preventive paradigm.

The majority of Internet network traffic exists within the confines of network services and protocols, and their corresponding ports. The list of ports in the opening exercise of this chapter is just a minute sampling of all the possible services allowed in the Internet via the associated ports. In networking parlance, a port is defined as endpoints of logical network connections, established for extended communication purposes, between two interconnected network devices exchanging data via an established network service. The port numbers are divided into three ranges: the Well Known Ports (0-1023), the Registered Ports (1024-49151), and the Dynamic or Private Ports (49152-65535).

The ports listed at the beginning of this section correspond to those associated with major intrusion attempts, and the line between an intrusion attempt that has been properly neutralized and an intrusion attempt that culminates in an intrusion incident is a very thin line; the difference resides in the situational awareness of the targeted entity. If this entity monitors its inbound and outbound network traffic from a proactive posture, an incipient attack will be promptly detected and properly neutralized. Else, the targeted entity will become yet another statistical victim. Let us explore a basic enumeration of the listed ports:

The strange and anomalous port number "0" is defined by IANA as "reserved" (both TCP and UDP). The documentation on RFC675 states that a socket of all zero is called unspecified, and the purpose behind unspecified sockets is to provide a sort of "general delivery" facility, thus resulting in the socket 0.0.0.0:0 or n.n.n.n:0, where "n" is a valid octet value. However, there's no valid reason for using this socket with port 0, and the presence of this peculiar socket should be considered suspicious at the very least.

Network traffic on ports below 20 (and 37), commonly known as the "small services" can be safely blocked at the gateway, since these tasks are primarily system management, and they shouldn't appear on inbound traffic, unless specifically authorized for approved IP addresses, though under normal network operations these services are not needed.

There's an interesting attack developed years ago using the echo and chargen ports, by injecting packets with spoofed source and destination, thus misusing the normal responses of both the echo and chargen services to generate a DoS attack. Cisco's routers can enable/disable the services echo, discard, and chargen.

Since its initial proposal in 1971, FTP is one of the oldest protocols, designed to facilitate the sharing of files and the reliable transfer of data. After the client and the server negotiate the data transfer parameters via port 21, the FTP server opens a second connection for data on port 20 to the client. This protocol is insecure for data transfer since everything is sent in the clear where user names, passwords, FTP commands and transferred files can be captured using a packet sniffer. There are safer alternatives for data transfer[61] and the use of FTP should be disallowed.

---

[61] See SSH section ahead

Secure Shell (SSH) is commonly deployed on port 22/TCP, and it has become the de facto connection schema for command line access on both Unix and Windows. SSH is a very popular target among attackers seeking to brute force root accounts. The first defensive measure (but not the only one) should begin by deploying SSH on an alternate port other than the default 22/TCP. Other defensive layers include the deployment of SSH brute force prevention tools,[62] disallowing remote root logins, and using SSH authentication keys instead of password authentication.

Port 25 is customarily used by mail servers to process SMTP[63] traffic, in itself a communication capability considered a mixed blessing, since many cyber attacks use SMTP as their preferred attack delivery tool, with spam levels rising as high as to a 90% range. SMTP wasn't the first "e-mail" protocol; it was preceded by UUCP[64] as an e-mail exchange protocol in the pre-GUI[65] era.

The e-mail exchange begins with an e-mail client signaling to the e-mail server its intention to connect, and the latter will respond with a "220" line, indicating it's ready to receive e-mail data, provided the client identifies itself with either a "HELO"[66] or an "EHLO"[67] line message.

If the initial exchange of line messages is successful, the e-mail server will respond with an acknowledgement line message "250" and then an established connection between client and server takes place. SMTP represents a serious security risk due to its lack of authentication mechanisms, and allows the injection of forged data in the SMTP headers, thus leading to and facilitating the abuse via spoofed e-mail sources and forged text data.

---

[62] DenyHosts, Blockhosts, etc.
[63] Simple Mail Transfer Protocol
[64] Unix-to-Unix-CoPy
[65] Graphical User Interface as popularized by Microsoft Windows applications
[66] According to the old RFC 821
[67] According to the modern RFC 1869

IPsec is a suite of protocols that together provide encapsulation, authentication and encryption of traffic, and it's a common solution for corporate VPN gateways, customarily operating from ports 50, 500, and 4500.

IKE (Internet Key Exchange)[68] serves as the most popular protocol used to authenticate a VPN session, and usually implemented on port 500/udp, using a Diffie-Hellman key agreement, and adopting either a "Main Mode" or an "Aggressive Mode" for this process. There is a security weakness when in Aggressive mode, since some data is exchanged before a secure tunnel is in place, thus facilitating a sniffing attack resulting in a compromised connection.

ESP (Encapsulating Security Payload) is the most common protocol for encapsulation of the actual data in the VPN session. Since ESP is IP Protocol 50, it's not based on TCP or UDP protocols, thus creating problems for NAT devices, often experiencing problems with ESP. If an intermediate device passes only TCP and UDP traffic, the ESP encapsulation will fail, since ESP is IP Protocol 50, i.e., it's neither TCP nor UDP. NAT-T-IKE allows the ESP session to be encapsulated within a more NAT-friendly UDP packet on ports 500 or 4500 UDP.

Domain Name Service (DNS) is the de facto client/server protocol allowing clients to resolve hostnames into IP addresses (and vice-versa). This protocol is one of the favorite targets for cyber threats and attacks. When a client requests the resolution of a hostname to its corresponding IP address, the client sends a UDP port 53 packet to its assigned DNS server. The DNS server looks into its own table, and if the requested resolution is not found the query is forwarded to other DNS servers on the Internet. The distributed DNS database is hierarchical and it's based on 13 root DNS servers.

DNS servers have "zone files" that contain the hostname to IP address tables, and DNS servers are customarily setup in a

---

[68] Formerly known as ISAKMP (Internet Security Association and Key Management Protocol)

master/slave relationship, thus enabling the slave servers to download the zone files from the master at configured intervals. The zone transfer occurs over TCP port 53, for ensured reliability of the transfer data. The original DNS query is done via UDP for the sake of expediency.

Perhaps the most insidious DNS attack is cache poisoning, designed to alter the IP/hostname table in a DNS server's cache, with disastrous results. The only effective way to prevent DNS cache poisoning attacks is by implementing DNSSEC.[69] A successful cache poison attack is difficult to detect, and can effectively paralyzed the targeted organization.

When an attacker gain access to monitor the DNS queries he can also forge responses, thus seizing control of the IP addresses returned on the DNS responses. Until DNSSEC is fully implemented, these attacks remain a serious threat to any organization.

The very usefulness of the Dynamic Host Configuration Protocol (DHCP) represents its mayor weakness, because is ubiquitous in any TCP/IP network. This protocol allows for the automatic assignment of TCI/IP configuration, from a DHCP server to a host, thus enabling it to become a member of a local network, for the duration specified in the IP address lease period. The traffic between the DHCP client and the DHCP server occurs on ports 67 and 68 UDP,[70] respectively. Precisely because of its use of UDP, DHCP datagrams are easily spoofed, and IP leasing exhaustion can easily lead to a DoS attack. Furthermore, rogue DHCP servers can effectively disable all clients on a local network

And so we have arrived at the summit of all threats, represented by traffic on ports 80 and 443! The former is the port for the WWW traffic (HTTP), and the latter is the secure iteration of the same,

---

[69] An extension to the original DNS protocol providing digital signatures of the DNS zone files.
[70] User Datagram Protocol, a stateless and connectionless protocol void of reliability and integrity mechanisms.

namely, HTTP over TLS/SSL, or HTTPS. Since both of them are required for WWW traffic, they are generally open through any type of border filtering device, allowing outbound and inbound network traffic. As such, these two ports represent a great threat to any enterprise because of the open policy associated with them.

However, the biggest mistake associated with the implementation of ports 80 and 443 is to equate openness with permissiveness, and to tolerate any type of content on them simply because of their association with WWW traffic. It is a necessity to keep them open, but it is a mistake to exclude this WWW traffic from close monitoring. The fallacy of considering this traffic as benign or neutral has paved the way to a plethora of cyber intrusions, and too many indolent policies leave this WWW traffic unattended because of its sheer volume. WWW browser attacks represent perhaps the highest percent of threats, both from insiders and outsiders. The challenge of monitoring WWW traffic is certainly great, but it's certainly not insurmountable. There is no standardized recipe for monitoring WWW traffic, but every enterprise with a proactive and qualified cyber defense team can tailor a monitoring strategy for their particular network configuration. The main point is not to leave the WWW traffic unattended, but submit this traffic under the analysis of qualified cyber professionals, who can design and implement trend and anomalous analysis on ports 80 and 443. It's not an easy threat mitigation solution, but it's certainly the responsible alternative to allow unrestrictive WWW traffic.

I know that this may become as a shocker to Windows users, but there are other OS besides the one popularized by Redmond.[71] Once upon a time, there was Unix, elder[72] over Windows by more than a decade, and still the favorite multi-tasking, multi-user OS among computer power users.

In a Unix server-client environment there are services using RPC, such as Network File System (NFS) daemons, and portmapper

---

[71]   The metonym for Microsoft
[72]   Developed in 1969

and rpcbind on port 111, both UDP and TCP. Portmapper is responsible for mapping services ports in Unix systems, by enabling the binding of remote programs. However, in some versions of Unix rpcbind may listen on both port 111 and ports above 32770. Therefore, implementing filtering policies that monitor only port 111 may not stop RPC traffic above port 32770, and an adversary may still be able to launch a compromise via RPC traffic on these higher ports. There are numerous vulnerabilities and exploits targeting services such as rpcbind and rpcmountd, and NFS, which could allow a remote attacker to gain root privileges and execute arbitrary commands.

One of the biggest gaps in network security is commonly represented by the use of the Simple Network Management Protocol (SNMP), with an abundance of devices[73] configured to use it, and the majority of them set to the default community string "public". With this community string in hand, it's just a trivial matter to query the devices using SNMP and elicit their configuration facts, or simply reconfigure them altogether.

SNMP is used to exchange management information between the SNMP management nodes and their corresponding agents installed in network devices. The SNMP manager polls the agents at periodic intervals over port 161 UDP, and the agent replies to the manager on port 162 UDP. This polling is done in conjunction with queries to the Management Information Bases (MIB) for the device in question. The MIB is a hierarchical database structure with information on the device (serial number, location, statistics, etc).

The threat represented by using SNMP is the information (traps) transmitted in clear text, and as such can be intercepted and manipulated. Of the three versions of SNMP (v1, v2c, v3) only the latter introduces encryption and message integrity, but all three versions are vulnerable to password guessing attacks. SNMP can use either UDP or TCP, but when using the former, packets can be spoofed and amplified, thus originating a DoS on

---

[73] Servers, workstations, routers, switches, printers, etc.

the SNMP-TRAP server, when exposed to the effects of being overwhelmed by thousands of traps per second.

The core cyber security issue with SNMP is the careful monitoring of the traps. When we allow SNMP traffic to transcend the boundaries of a local network, and into the Internet, the leaking of information may be potentially devastating, since SNMP can produce a network topology map, and in conjunction with the misuse of public community strings, will reveal the topology of an enterprise network to rapacious adversaries.

All interconnected networks sharing a homogeneous routing policy form an Autonomous System (AS). When an AS needs to communicate with another AS, they exchange data via the Border Gateway Protocol (BGP). Since this routing protocol, as many others, was designed before the Internet became a contested environment, it doesn't have built-in security, and therefore the information transmitted between routers using BGP is vulnerable to numerous attacks designed to alter or destroy the exchanged data. An adversary can insert malicious data into BGP and caused great disruptions between AS entities.

BGP operates over TCP, on port 179, and as such is exposed to all attacks targeting TCP. For example, a SYN flooding attack will cause a devastating DoS attack against the targeted router(s).[74]

On any given day the background noise on the Internet will be comprised of a significant percentage of traffic on port 445 TCP, always present among the most scanned ports, and one of the preferred ports used by botnets to scan and target vulnerable systems. And let's not forget the epidemic infection brought by the infamous W32/Conficker, Sasser, and Nimda malware cases, all of them using port 445.

The Windows OS provides file and printer sharing support through the Server Message Block (SMB) protocol on port 445 TCP, and

---

[74] Please consult RFC 4272 for additional details on port 179 vulnerabilities

while blocking traffic on port 445 at the firewall is a basic cyber security rule, many enterprises remain oblivious to this threat by failing to apply this security measure at their network perimeter.

Since its development in 1979, the Modbus Protocol is used to support client-server communication between ICS devices. From its origins as a proprietary protocol, it migrated to TCP/IP in 1999, simply to satisfy the convenience of users, and at the expenses of cyber security. Modbus has no security in its design, since it was intended to operate in closed networks, and attacking the devices controlled with Modbus is trivial. If someone can connect to ICS devices via Modbus on port 502, then the devices will accept any command. The rest is a simple extrapolation; an adversary connecting to ICS devices on port 502 can easily gain control on the targeted devices, and the more critical the function of the targeted device, the greater the negative impact on the ICS infrastructure.

Ports 1433 and 1434 are typically associated with network traffic for database connections. Since databases interact with local network applications, it presents a serious risk to allow ports 1433 and 1434 to be reachable from the Internet. However, numerous users of MSSQL leave these ports accessible from the Internet, and exposed to exploits. Malicious cyber predators repeatedly gained access into enterprise networks via these two ports.

Another database risk is present on port 1521 TCP, the default TNS[75] listener for Oracle, the popular relational database. Port 1521 should be properly secured, to avoid unauthorized and damaging commands to the listener, and to avoid unauthorized disclosure of database contents. The same precautions discussed under ports 1433 and 1434 (above) are relevant for port 1521.

The deployment of a proxy server on any given enterprise is a must, since it allows the management and control of Internet traffic in accordance with enterprise policies, and typically operating on TCP ports 80, 8080, or 3128. However, a misconfigured proxy

---

[75] Transparent Network Substrate

server may become an open proxy, thus creating the opportunity for cyber abuse by allowing Internet users to access your enterprise proxy.

Excessive use of your proxy server by Internet users is quite often an indicator of an intrusion, or the presence of malware facilitating anonymous browsing to unauthorized users, and in some cases may create liability for the affected enterprise. A vigilant enterprise should maintain a close monitoring policy on the traffic and usage of their proxy in order to avoid these types of misuse.

Over time the popularity of Redmond's Remote Desktop Protocol (RDP) has elevated this service as the de facto standard for remote administration in most enterprises, and the favorite of numerous datacenters, holding to the perception that RDP offers a secure data delivery, given its encryption settings.

However, the problem with RDP resides on the certificate used for encryption. Any default installation of RDP is vulnerable to a Man-in-the-Middle (MITM) attack,[76] whereby the encrypted exchange of RDP information can be intercepted and substituted, and without any warning to the affected parties.[77] The problems plaguing RDP are still present, and close monitoring of RDP port 3389 remains a must for enterprises that rely heavily of this service.[78] The RDP protocol can be made secure after applying the available hardening best security practices, but a default installation is vulnerable.

The Internet Relay Chat (IRC) is a text-based protocol designed to support a network of global nodes (clients and servers) forming

---

[76] First reported on 2003 on an article on MITM at http://www.securityfocus.com/archive/1/317244/30/0/threaded

[77] After Microsoft reported a fix for the problem exposed in 2003, a new article in 2005 reported the MITM vulnerability was still possible, at http://www.oxid.it/downloads/rdp-gbu.pdf

[78] Two Microsoft Bulletins published in May and August 2011 acknowledged the potential threats of Remote Code Execution and DoS affecting RDP

"channels" where members of the same channel can exchange messages. IRC traffic is typically found on ports 6667, 6668, 6669, and 7000, but it can really operate on multiple other ports.

IRC is a legitimate protocol that has become infamous as a result of its extensive abuse among the cyber criminals managing botnets, and deploying IRC as a command and control (C2) infrastructure. This malicious traffic is easily recognizable by the presence of commands and passwords, as opposed to the regular text conversation among the nodes members of an IRC channel. The presence of these C2 commands indicates the activity of a compromised client, now acting as member of a botnet under the control of a cyber criminal.

Perhaps the recent unveiling of the Stuxnet malware illustrates best the importance of maintaining a proactive monitoring cyber security posture based on trend and anomaly analysis, derived from a strategy of detecting anomalous behavior. Of course, this strategy must be based on the leadership and experience of qualified cyber SMEs, able to discern and analyze emerging network traffic trends exhibiting an anomalous behavior, such as the increase in Remote Procedure Call (RPC) protocol traffic and DNS queries to the unusual domains "mypremierfutbol.com" and "todaysfutbol.com".

Stuxnet makes extensive use of Peer-to-Peer (P2P) as part of the clever design of its developers, who, anticipating the potential unavailability of their C2 servers, built in a P2P update function in preparation for such eventuality. The P2P module installs an RPC server and client, and when a computer is compromised by Stuxnet, it activates the RPC server and listens for connections. Additional infected systems on the same network can initiate a connection to the RPC server, poll information on the installed Stuxnet version, and begin a P2P update process. This design circumvents the loss of the C2 infrastructure, and maintains all infected systems on the latest available version of the threat.[79]

---

[79] http://www.symantec.com/connect/blogs/stuxnet-p2p-component

This update P2P process generates an increase in network traffic for the RPC protocol when the threat C2 servers are unavailable, but when available, all infected systems will contact their C2 servers on port 80. This increase in HTTP traffic has been observed to focus on two domains, namely, "mypremierfutbol.com" and "todaysfutbol.com".[80] This HTTP traffic can only occur after the DNS resolution of these two domains has been completed, thus generating an increase in DNS queries to resolve these two domains. Any experienced cyber analyst will question these DNS queries as anomalous, because an ICS enterprise certainly has no legitimate reason to generate these types of DNS queries, since both domains are commercial and dealing with a topic completely unrelated to an ICS environment. Therefore, these anomalous increases in traffic, either on the RPC protocol or on DNS traffic to the two questionable domains, should raise a red flag in any trend and anomalous analysis on the ICS enterprise network traffic. This illustrates how proactive monitoring and analysis of network traffic, implemented and executed by perceptive network defenders, can and will take the initiative and nullify the offensive activities of a clever adversary. Nothing is hidden in network traffic; all it takes to unveil nascent threats is to know what is normal and what is abnormal, what is congruent and what is incongruent. If we delegate our defenses completely to automated systems, while we bask in indolent behavior, expecting a magic alarm to indicate the arrival of a cyber threat, we will be for ever living in defeat, knowingly or unknowingly.

This same Stuxnet case illustrates best the general ignorance pervading organizations, both great and small, and leading them to follow the speculations posted by media reporters on articles meant to sensationalize the issue. Without a thread of forensic evidence the media has escalated the origin and scope of the threat to heights standing in blatant contradiction with the available[81] forensic evidence. Wild theories are built on the potential

---

[80]   Ibid; W32.Stuxnet Dossier, Version 1.4 (February 2011), Nicolas Falliere, Liam O Murchu, and Eric Chien

[81]   As of the end of September 2010, when this section was being written

target of Stuxnet, and consequently, on the alleged identity of the malware's developers, with total disregard for the available statistical evidence, which lacks support for such outrageous speculations. Let us remember what we already discussed in previous chapters. A cyber incident cannot be oversimplified and reduced to a linear event model. Cyber incidents are usually developing in a nonlinear fashion, lacking a direct line between root cause and effect.

To ascertain the possible origin and target of a distributed malware a collection of forensic data must be obtained and analyzed. As of September 2010 Stuxnet has targeted the critical infrastructure of several countries, thus contradicting the media speculation supporting the theory of a single target. Both commercial and US government organizations ascribe an undeserving degree of credence to the media speculations regarding the origin of the malware, simply because of the generalized degree of ignorance regarding the radical nature of cyber attacks. The two determining factors when seeking to establish attribution of a cyber threat are potential indicators in the source code and/or HUMINT. In the absence of these two elements, any attribution is mere speculation.

As previously[82] stated, the unprecedented fact is that in our perennial state of asymmetric cyber conflict, generated by the tremendous disparity between those who know and those who don't know, cyber knowledge becomes the ultimate equalizer. Any person with in-depth cyber knowledge and programming skills can bring the entire world to its knees. This person no longer needs an army, or a national treasure, or transportation logistics to support a cyber attack. Cyber survival and cyber superiority depend only on cyber knowledge, cyber vigilance and the complex CYBINT paradigm. However, this profound concept is not generally implemented, because in our cyber ignorance we turn to automated systems, and we entrust our cyber security to them. Thus, we failed every single time in our quest for protection, because we think that these cyber appliances will save the day. This fallacy thrusts us even

---

[82] Chapter IV

deeper in our weak condition of cyber security disparity in the asymmetric conflict. I'll repeat what I said before:

> A cyber conflict is a battle of brains, not of machines

A cyber attack can only be detected, understood, defused and counteracted after collecting, examining and analyzing the cyber attack dissemination methodology, the attack vector, and the associated cyber code of the malware in question, when and if present. A cyber attack does not always require the presence of malware, but it always requires a cyber attack vector.

Chapter 12. Cyber intrusion

Why waiting until now to speak about this topic, which could have been addressed at the beginning of this book? Because a cyber intrusion is much more than a technical issue, and a vast array of factors and circumstances surround a cyber intrusion event. The previous topics, then, have hopefully prepared the reader to obtain the full benefit of this chapter.

Technically speaking, a cyber intrusion takes place when an unauthorized set of instructions interact with a cyber code installed on a system selected as a target. The unauthorized interaction may or may not include the injection of unauthorized cyber code. In the case of injected code, this may not be necessarily a malware code, since the injected code may have a legitimate use, but it is being utilized without the knowledge and/or the authorization of the system's owner.

In cases when resident code is used in an unauthorized manner, the system's owner has knowingly or unknowingly allowed such resident code to remain active, and the intruder will use the corresponding active cyber service as the preliminary attack vector. In cases when code injection has taken place, the system's owner has facilitated the injection by allowing the presence of an unpatched cyber vulnerability, or by accepting interaction with the attacker via a deceptive lure. In all instances, the common factor is the system's owner performing or allowing an action leading to an intrusion. This fact underlines the paramount cyber security axiom; a cyber intrusion takes place when the owner of the victimized system allows the unauthorized activity to take place. The fact that the cyber intrusion may occur without the system's owner's awareness is irrelevant, since unawareness will not preclude the successful attack.

Is this a harsh assessment? Absolutely not! This is only a factual assessment emphasizing the responsibility and accountability of the system's owner. Establishing a cyber presence and interacting in cyber space is not a game, and every cyber intrusion has a cause pointing toward the system's owner. What is the corollary? If you own a cyber system, you should know how to protect it, and if you don't know how to protect it, seek assistance from reputable cyber professionals. Cyber space is not an amusement park, where you can ride and expect to be protected by those providing the entertainment while you simply enjoy the ride. In cyber space you need to look after your own preservation. There are numerous reputable organizations and accredited individuals ready to assist you, but you must be willing and prepared to do your own homework in achieving this self-preservation goal.

Intrusions in general can be categorized in two main groups: predictable and unpredictable, and up to 95% of all cyber intrusions fall within the predictable category, while a very small percentage truly belong to the unpredictable category, such as those caused by zero-day vulnerabilities. The predictability of cyber intrusions is a fact systematically denied by the ubiquitous presence of those who sponsor the imaginary existence of an über-cyber species capable of a mythic degree of cyber "sophistication"; the buzz term preferred by the uninitiated who want to sound as if they understand cyber intrusion dynamics, even though they don't. The abuse of this term, either in its adjective or noun form, is the buzz term used by the media, and regurgitated by those who follow the media, whenever they attempt to discuss a cyber event that floats above their level of understanding, but allow them to use the euphemistic adjective "sophisticated" in order to avoid the more realistic terminology "I don't understand the technical details of this cyber event, but I don't want to sound ignorant." The "sophisticated" euphemism is also used and abused by those who prefer to avoid personal and organizational accountability when dealing with a cyber incident caused by their ineptitude and lack of cyber security posture. How can anyone possibly be blamed for a cyber intrusion when battling against an über-cyber species so "sophisticated" that no one could possibly defeat?

I'll reiterate what I said before. Any cyber conflict is a battle of brains. The victor is always the one with superior cyber knowledge, and this cyber knowledge is available to anyone. Therefore, any cyber user can and should acquire the required cyber knowledge necessary to protect the cyber system for which one is responsible. This is the foundational principle in this book; cyber is truly the ultimate equalizer. If you work hard to learn and continually study the cyber principles, then you will not only stand your ground during a cyber conflict, but you will also become the victor. In cyber there are no über-cyber species; only über-cyber knowledge! The more you know, the more powerful a cyber entity you become in the cyber environment. Conversely, there are no innocent cyber victims, but rather uninformed cyber victims. Am I suggesting that every cyber user should become a cyber expert? Absolutely not! The premise of this book is to promote the notion that anyone can and should be capable of defending one's own cyber system, or taking the initiative of seeking assistance from cyber experts, at whatever level such expertise may be available in the area where the user resides.

Every user should abandon the attitude of utter despair, assuming there is no possible defense, and begin a new chapter in their cyber life, by rejecting defeat. In cyber there is no secret knowledge enabling the mythical über-cyber species. In cyber all the knowledge is in the open, available to anyone who seeks this knowledge. Users should not see themselves as powerless prey in the paws of cyber predators. Users should see themselves as defenders of their private cyber kingdom, and gain the knowledge that will empower them to resist and succeed in defending their cyber castle. The good news is this: while the cyber code written to create a new cyber application contains code weaknesses, there is always a countermeasure available to oppose and defeat the adversary's attack. In cyber there is no invincible opponent, because cyber is the ultimate equalizer.

In the previous chapter we did a cursory review of the impact associated with the different ports, and their corresponding services, open on any particular system on a network. In this chapter we also review the "foreign" concept of cyber accountability. It appears as if we live in this dysfunctional universe where we have ownership of cyber systems, but we never take responsibility for the cyber intrusions against the same systems, and all the mistakes, by either omission or ignorance, have no consequences on the owners, who continue repeating the same errors ad nauseam. The same aversion of accountability is leading to depend more and more on automated system designed to do our job, by becoming systems configured to protect other systems. We want to escape into this utopian world where we can use all the cyber systems we want, but take no responsibility for their cyber security.

In this context is it prudent to revisit once more the concept of cyber accountability sponsored by this author. How much are we contributing to the opponent's success when we failed to patch applications that are ubiquitous in our AOR, such as PDF readers, Microsoft Office, Java, Flash Player and web browsers? To what degree are we negligent in failing to patch OS in our servers and clients, IOS in our routers, and iOS in our mobile systems, exposed to specific vulnerabilities? How many times do we fail in minimizing the number of users with administrative privileges, thus exposing our AOR to internal threats, or to the cyber vandalism of former disgruntled employees?

The proactive and intelligent cyber defense advocated by this author is an achievable and affordable reality, not the "Hollywood fantasy" perceived by the non-enlightened minds of the cyber pretenders that the only reality they know is that of cyber defeatism.

There are proponents of complete cyber automation that envision a strategy based on an arsenal of automated cyber tools that are rarely deployed because of the risk of disclosure, fearing that once used and discovered by their opponent they disappear from their active arsenal. This is very representative of the reality of "canned" cyber tools, which are designed to operate in the vacuum of qualified cyber experts. Cyber tools designed by cyber experts are not limited to the category of single-shot tools, since they are tailored for a particular cyber encounter, and adopted to the particular dynamics of such an encounter. As such, they can be reuse by tailoring them to the new set of dynamics germane to a new cyber encounter, where the different cyber content will mutate the cyber tool into a different entity. Canned cyber tools are the result of engaging in a cyber encounter with a preconceived cyber arsenal that sacrifices the role of human brain in the altar of cyber automation, and this choice is done in the absence of qualified cyber strategists and cyber designers, working in conjunction with CYBINT.

In the content of a cyber encounter generating an exchange of cyber intrusions there must be a minimum of two opposing parties, each one with a peculiar cyber footprint as their point of presence in the Internet. When qualified cyber strategists, cyber designers and CYBINT assets work in synergy to create a target tool designed to affect the opponent, there is no need for canned cyber tools. As the encounter is dynamic and ever evolving, so is the cyber footprint of each opponent. The cyber tools designed to create a particular effect on the opponent are therefore, by necessity, also dynamic and evolving. Those who sponsor the theory of a static arsenal of cyber tools designed for a single shot effect are unaware that every cyber encounter is highly complex and nonlinear, and single shot linear tools are inadequate to achieve victory over the opponent.

Let us reiterate what we previously discussed in chapter 2 regarding the importance of defining the specific environment where the conflict takes place.

Not everything electronic is cyber, and the electromagnetic spectrum and cyber spectrum are two different realms. Those who blur and attempt to fuse these terms do it in ignorance and in doing so they mud the waters in our search for the right answers.

During a cyber confrontation the opponent who knows his cyber environment, and studies his adversary's cyber environment, is the one with the best opportunity for success. The research and the study of one's own cyber environment, as well as the opponent's, begins and ends with the search and study of the binary data collected during the cyber confrontation. This is where cyber forensic research becomes the foundation for cyber victory, but cyber forensics can only be conducted by human beings, not by automated systems.

This author recently gathered cyber forensic data showing an anomalous network internal activity performed by an unauthorized User 2 on the computer of a legitimate User 1. The latter logged into his assigned enterprise computer and performed his daily activities of reviewing and writing documents residing on his hard drive. User 1 did not initiate any browser activities, but an examination of the raw network data captured with Wireshark[83] revealed that User 2 established a remote connection into User 1's computer and from there initiated a browser session detected by Wireshark as an HTTP POST command. User 2 is a former member of the group employing User 1.

This is a cyber intrusion that cannot be detected with automated tools, but only through the collection and study of the cyber forensic data made available with network monitoring tools, deployed as part of conducting and maintaining a proactive monitoring program depending on the presence and support of qualified cyber professionals. This presence and support is the one factor that provides the advantage over completely automated cyber tools, unable to think outside the box, and therefore unable to detect these types of incidents. Yes, we can benefit from the synergy between automated cyber tools and qualified cyber professionals, but complete and exclusive reliance on the former is the secure recipe for a cyber defeat.

Let us apply the concept of cyber accountability to this internal incident. How did User 2 acquire the necessary permissions to execute a remote login onto the computer assigned to User 1? Were these permissions part of User 2's assigned duties, or is

---

[83] A reputable and widely use network analyzer used by qualified cyber professionals. See chapter 4 sample of Wireshark captured data.

this a case of excessive permissions granted to a user? And when User 2 left the group where User 1 belongs, why did User 2 retain those network permissions?

One of the cardinal rules in cyber security is the principle of least privilege, which translates into assigning to a user account only the very essential privileges required for that user to accomplish his assigned work. Records associated with User 2's account indicate a violation of the principle of least privilege, since in his assigned work he didn't need permission to execute remote login. A second violation occurred when User 2 moved to a different location and to a different job description, but his privileges for remote login were neither revoked nor disabled. This is, in sum, a case when a cyber intrusion is facilitated from within the same organization where the intrusion takes place.

We must put an end to the perennial state of cyber innocence in which we have become accustomed to live, and we must replace it with a mindset characterized by a healthy measure of caution and alertness. In this transitional state we don't have to become experts in cyber intrusion, but we must strive in developing that sense of alertness that generates indicators when something unusual and abnormal affects our networked cyber assets.

We must abandon the overly simplistic concept that cyber defense means having an AV application installed in our computers. Cyber intrusions have a plethora of attack vectors beyond the narrow area of protection provided by AV, and even that narrow area of protection becomes effective when, and only when, such AV is reputable, dependable, and current.

A cyber intrusion is not always a disastrous event, but it is always an invasive event. Do we know what our computers are doing? Are we always in control of the activities of our computers, or are there overt cyber activities initiated by unknown or unmonitored cyber services activated in our computers beyond our level of cognizance?

Are we unwillingly participating in malicious cyber activities because our computers are no longer under our exclusive control? Are we as owners monitoring the activities of our computers, to ensure that their activities are always under our control as their legitimate owners?

> We want to live in this dysfunctional universe where we have ownership of cyber systems, but we never take responsibility for the cyber intrusions against the same systems, and all the mistakes, either by omission or ignorance, have no consequences on us the owners, as we continue repeating the same errors ad nauseam.

How many cyber intrusions go unnoticed by those who entered the cyber arena via the metamorphosis process described in chapter 4, because they lack the skills set required to detect insidious cyber intrusions? There is a plethora of reports and statistics on cyber intrusion that only acknowledge "documented" cyber intrusions recorded by automated cyber detection systems. These statistics are grossly underestimating the real number of cyber intrusions that remain unrecognized by automated cyber detection systems. In turn, these faulty statistics contribute to the creation and dissemination of skewed cyber reports that provide leadership with a false sense of security, thinking that their cyber security posture is not as bad as they thought.

This author has worked in enterprises where the official cyber reports were limited to count the "documented" cyber incidents, as compiled by the statistics of those incidents that were detected by automated systems, and recorded only if they exceeded the threshold of a certain amount of outbound data related to that same incident. The actual number of cyber incidents in that enterprise was always higher than the number included in the official report, but those additional incidents were ignored because they were detected by experienced cyber SMEs, and they did not complied with the requirement of meeting a threshold of outbound data. This enterprise worked on the misunderstanding that only a cyber incident accompanied with outbound data constituted a true cyber incident.

By contrast, most of the cyber incidents discovered by the cyber SMEs constituted serious threats to the network of this enterprise, because such cyber incidents were cases of DoS, DDoS, SQLIA, and buffer overflow intrusions. None of these types of intrusions were detected by the automated systems, and none of them generated outbound data to trigger the official thresholds.

This contrast brings into focus a very important issue in the topic of cyber intrusions. How do we measure them? How do we report them? Every enterprise must find a systematic, logical, and financially feasible answer to these questions, because any cyber defense must exist within the affordable settings dictated by the cost of doing business. But we must also ask ourselves this: What is the cost of sustaining a false sense of cyber security because we don't want to face the real facts concerning our cyber posture?

Why do we discuss the topic of cyber intrusion as a separate chapter from the topic of cyber attack? Because in the strategy of a cyber conflict, a cyber intrusion typically represents a prelude to a cyber attack, and the former may simply represent the insertion of the adversarial cyber code into the targeted network, where it may remain dormant, and often undetected, until the activation of the adversarial code, thus initiating the cyber attack stage, generating cyber activity that will modify the normal cyber operation of the targeted systems.

The well publicized Stuxnet case illustrates this distinction quite clearly. A study of the timeline of the Stuxnet cyber intrusions, based on reputable and qualified forensic evidence, reveals that the time span between the cyber intrusion and the discovery of the impact of the consequent cyber attack took over several months. The public disclosure of the Stuxnet infection was originally reported by VirusBlokAda on 17 June, 2010, but Siemens did not publicly acknowledge the impact of Stuxnet until 14 July of that year. How long before the Stuxnet code activation the insertion took place and planted on the affected systems? The time stamp forensically recovered from some of the main Stuxnet components revealed that such files were compiled as early as January 2009.

Consequently, we can only retroactively establish a timeline for the Stuxnet infection that spans over a year, and yet, we remained publicly unaware of it during that time span.

Describing the insertion of adversarial cyber code as an often undetected prelude does not imply that a cyber intrusion is undetectable. It typically remains undetected by the deficient cyber security practices that rely exclusively on automated detection systems. However, a cyber intrusion is perfectly detected by qualified cyber SMEs who maintain a vigilant eye over the targeted network. The axiom sponsored by this author rings true once again: know your network, and you will know when a prelude to a cyber attack is in progress.

Knowing the network under our care means to know the kind of network activity that should be considered normal and authorized. The concept of authorized activity is based on an in-depth knowledge of cyber security policies, and the enforcement and compliance to such security policies. The unavoidable vulnerabilities in any network should be constantly assessed and monitored, in order to create a cyber activity baseline that serves as a benchmark. This baseline is critical to our cyber security posture, and the ever evolving network dynamics of our network can always be compared to this baseline, in order to determine if any deviation corresponds to an unauthorized or unknown network activity. This knowledge is the difference between detecting cyber intrusions, and being targeted by cyber intrusions that go undetected.

Knowledge and diligence in detecting a cyber intrusion are a priority, since the insertion of adversarial binary code does not necessarily remains dormant for an extended period of time, as in the cases of DoS, DDoS, SQLIA, and buffer overflow intrusions. And yet, with the proper knowledge, diligence and proactive cyber defense operation led by experienced and qualified cyber SMEs, even swift cyber intrusions can be detected in time to nullify the adversarial activity.

Chapter 13. Cyber attack

One of the most misunderstood concepts is what constitutes a cyber attack. Both the press and government officials have written thousands of pages on articles and reports concerning cyber attacks, and through the dissemination of these documents they continue perpetuating the misunderstanding on the essence of a cyber attack, in a never ending vicious circle of misnomers.

Any online search on the topic of cyber attack will generate a plethora of articles written by both industry and government representatives, but this abundance of articles shared a common denominator: a conspicuous lack of the definition of what constitutes a cyber attack. Instead, they undertake the task of speaking about all the different types of cyber attacks, while neglecting to define the very essence of the topic of their writing.[84] Once again, I'd like to point out the inadequacy of writing about a subject matter while neglecting to define the very same subject matter. This symptom truly proves one of the thesis of this book, namely, many seek to become spokespersons for a subject matter they do not know how to define.

There is only one type of attack vector in a cyber conflict, and that is a piece of binary code designed to modify or terminate the execution of the original code as designed by the original author, and without the consent of this legitimate author. The modification may occur in direct interaction with the original code residing on a target system, as an injection of foreign code into

---

[84] As a random example of this problem, a reputable technical magazine published an article in 2010 where the very question of what constitutes a cyber attack is introduced, but left unanswered. Instead, the writer proceeds to explain a DDoS as one of "the most typical" http://www.pcmag.com/article2/0,2817,2358610,00.asp

the target system, or as an unauthorized execution of resident code. Anything else beyond this description does not qualify as a cyber attack. Consequently, those who speak and write about electronic warfare as constituting a cyber attack are simply misunderstanding the very essence of a cyber attack.

The missile or the explosive device delivered by any weapons platform, whether operating from air, sea, or space, is the attack vector in any conventional warfare conflict. However, the carrier platform is not the weapon, but simply the delivery device. Likewise, a cyber attack vector, namely, a piece of cyber binary code, may be delivered by a number of different carriers, such as radio waves for example, but the carrier is not the cyber attack vector, because the carrier is not a cyber element, but only its conveyance.

Communications based on radio wave technologies are ubiquitous in our current society, and systematically misunderstood and misidentified by those who misplace everything that is based on electronics as part of the cyber medium. The cyber medium is unique, and it's comprised of only one singular component; the binary cyber code. This cyber code may be delivered by utilizing many of the different electronic communication technologies, but they work only as a conveyance for the cyber code. Not even a network infrastructure is a cyber attack vector in itself; only the binary cyber code carried by that network is the cyber attack vector.

Regretfully, this very important distinction is almost inexistent in the collective conscience of those who attempt to lecture and pontificate about cyber attacks. Radar or GPS jamming devices do not constitute a cyber attack vector, simply because there is no binary cyber code in radio waves. And even when radio waves are used to carry and deliver a cyber code, the attack vector is the binary cyber code itself, not the conveyance.

An attack against the cyber code conveyance will certainly have a negative impact on the operational aspect of the cyber code, but not on the integrity of the cyber code itself. Since the cyber code conveyance may be obstructed, thus affecting the delivery of the cyber code, the intended execution of the obstructed code will remain in suspension until the obstruction is nullified. However, this suspension neither terminates the execution of the obstructed cyber code, nor modifies the cyber code, and therefore, this scenario does not constitute a cyber attack. The important task in this scenario is to identify the precise nature of the obstruction, and design an effective counter measure to eliminate the obstruction.

Since the obstructed cyber code remains unaltered, the obstruction is not a cyber attack, but quite likely an EMS attack. In this case, EMS specialists should analyzed the obstruction and counteract it, so as to remove it or bypass it, thus rendering the cyber code operational again.

This scenario illustrates how important is to establish a clear distinction between an attack on the cyber code, and an attack against an associated media used as a conveyance for the cyber code. If we failed to establish this difference we will fail in implementing a timely and effective countermeasure as well.

The importance of this distinction transcends semantics, because a proper definition of what constitutes a cyber attack is crucial in defining what constitutes a cyber conflict. It is critical to have a sound definition of a cyber conflict because such definition has vast and crucial ramifications into the realm of policy making, international laws, and national security.

A cyber attack vector is a piece of binary code designed to modify or terminate the execution of the original code as designed by the original author, and without the consent of this legitimate author.

Since the very essence of the cyber spectrum is the binary code, and every binary is written for a particular purpose and its corresponding tasks, it is simply impossible, both intellectually and pragmatically, to anticipate all the feasible ways by which such binary code may become vulnerable to an attack.

This situation was amply illustrated in chapter 11, where we discussed the many misuses of well-known protocols. The code for each one of these protocols was written for a very specific task, and to benefit Internet users. However, it didn't take long for the malicious mind of those who seek unlawful profit or the harming of innocent users, to find a way to subvert the original purpose of the legitimate code, and use those protocols for a nefarious purpose. Against this condition of the human race there is not technical solution. We are naturally predisposed to evil and selfishness. So the countermeasure must come from a change in our mind set, and assume that our network will always be targeted by adversaries, and thus we must exercise caution and assess our cyber weaknesses on a continuous basis.

We have mentioned before that risk management is perhaps the best possible solution to protect our cyber posture on the Internet. By the very simple fact of being connected to the Internet we become vulnerable. The only question is: how vulnerable?

There are some very reputable organizations that can assist any AOR into assuming a proactive cyber defense posture, and maintaining it. The non-linear complexity of the cyber spectrum dictates that our cyber defense posture must be continually reassessed, because the vulnerability factor is in constant flux, and it requires constant attention. Those who are appointed to assess an enterprise cyber defense posture must be qualified and experienced cyber SMEs. The important mission of cyber defense assessment should not be placed in the unqualified hands of users adopted via metamorphosis, or even less, entrusted exclusively to automated systems. Cyber defense posture is a matter of brains, knowledge and experience, not a matter of machines. The following actual case illustrates the non-linear complexity of cyber incidents that should be handled by alert and analytical qualified cyber SMEs, and not entrusted to beginners assimilated by the metamorphosis process.

An important organization was victimized by a recurrent SQLIA series. According to the forensic evidence collected by this author, the initial attack took place on March 2006, but the cyber defense division responsible for protecting the targeted network failed to identify the attack, event though the initial intrusion was followed by several other SQLIA incidents during March, May, August, and September of the same year. The incident was eventually acknowledged on 28 September 2006 by the cyber defense division, seven months after the initial incident, and only after receiving the undeniable forensic evidence provided by this author.

When the negligent division was officially tasked to provide an explanation for their negligent behavior, their response stated they have modeled their analysis of SQLIA after a previous incident targeting a sister organization, showing an attack string associated with the "200 OK" web server response to the attack. They added that the SQLIA targeting them had not displayed the same response from their web server, and their weekend analyst (non-certified at that time) had deemed the attack as unsuccessful.

The damaging series of SQLIA is the result of employing unqualified and inexperienced personnel tasked to provide cyber defense services, relying on static models based on the fallacy that all SQLIA are the same, and expecting the same response from the victim web server. The serious impact caused by the SQLIA on the victim organization warrants at the very least a corrective action of retraining and certifying cyber analysts to recognize the methodology of an SQLIA. This type of attack is predicated on illegitimate user input and unauthorized "error messages" provided by the victim server. These two elements are crucial in the detection of SQLIA, and all available logs should be searched for the presence of these two elements. A mechanical and limited search for the "200 OK" response is incomplete and ineffective. A successful SQLIA detection is to be predicated on a comprehensive and analytical search, focused on the presence of illegitimate user input and unauthorized "error messages".

There is no linearity in cyber attacks, as illustrated on this case where the analysis of the pertinent logs shows the evidence of the successful SQLIA is indicated by a "500 message error". Cyber defense cannot depend exclusively on check lists followed by individuals lacking the knowledge and experience of a qualified cyber SME. The disastrous results suffered by this targeted organization led to Privacy Act data extraction, intrusions with escalation of privileges, email spoofing, and web defacements. Cyber defense requires a human brain molded and trained to operate in accordance to the skill set and mindset required by cyber operations.

There is an excellent cyber defense framework[85] that has received the endorsement of both government[86] and industry,[87] and it is publicly available. This framework is the result of balanced and in-depth study of the cyber defense issues affecting any enterprise connected to the Internet. In addition to the adoption of a framework to achieve a sustainable cyber defense posture, it is equally important to implement a risk management program that works in conjunction with such framework. The very nature of doing business or sustaining a mission on any enterprise network requires that we have a model that allow our enterprise to qualify and quantify the risks that can be accepted, and the ones that must be avoided in order to maintain operational status. The implementation, active monitoring, and sustainment of both the cyber defense framework and associated risk assessment plan are imperative in order to achieve a defensible cyber posture that is capable of providing a resilient cyber presence.

The aftermath of a cyber attack offers the best opportunity to reexamine our cyber security policies, the effectiveness of our cyber security controls, our risk assessment plan, and our cyber incident response plan. Since cyber incidents are unavoidable, due to underlying cyber code errors and the ever-present human

---

[85]   SANS 20 Critical Security Controls. Make sure you consult the most recent and updated one.

[86]   DoD, NSA, DHS, FBI, DoE, US CERT, among others.

[87]   Civilian penetration testers, among others.

errors, our cyber incident response plan (CIRP) must outline the criticality of the enterprise systems into various levels (Criticality 1 through Criticality n), where 1 represents systems that are deemed dispensable, and Category n represents systems that are indispensable for the enterprise mission.

During a cyber attack the only feasible and effective response is the identification of the systems that are vital to the survivability of the enterprise, and the assignment of the best available defenses to the systems categorized in that critical group. The incident response plan also must assign the higher level of authority to the cyber defense qualified SMEs leadership, so that all defense actions are coordinated with the different boundary system administrators under the orders of the Lead Cyber Defense SME (LCDS).

The CIRP must be activated as soon as the LCDS obtains and analyzes the cyber intrusion indicators announcing the impending and consequent cyber attack, notifies the enterprise leadership, and obtains authorization to activate the CIRP. The first course of action upon activating the CIRP is to place all boundary system administrators under the authority of the LCDS, and activate logs collection on all boundary systems, whenever feasible.[88] These logs will create the forensic evidence on the dynamics of the cyber attack and the corresponding cyber defense, creating an invaluable record that will allow the analysis of the cyber interaction, and providing a wealth of data that will enrich the process of reassessing the enterprise's cyber security policies, the effectiveness of our cyber security controls, our risk assessment plan, and our CIRP.

During a cyber attack the targeted victim (individual or enterprise) has the inalienable right to protect the network enclave under their ownership and care. Regretfully, many enterprises are organized under a centralized regional enclave that doesn't allow local

---

[88] Typically, most enterprises do not keep log collection active on their boundary systems during normal operations, due to storage and financial limitations.

networks to exercise this self-defense prerogative, since there are no administrator permissions granted at the local level. This model is responsible for allowing countless cyber attacks that remain undetected and unmitigated. The most effective model of cyber defense is based on the in-depth knowledge of the local network undergoing the cyber attack. A centralized model of cyber defense fails to comply with this crucial requirement, because an AOR exceeding the capabilities of in-depth knowledge achievable by remote personnel is counterproductive. The degree of coordination, synergy, expeditiousness, and in-depth knowledge of a network enclave at the local level is not a scalable property. Only local network administrators and local qualified cyber defenders, with profound knowledge of the topology and behavior of their local network can feasibly aspire to mount a cyber countermeasure defensive plan with realistic possibilities of success.

In the case of an individual willing and able to protect its home network against a cyber attack, the most effective cyber defense plan may depend on the feasibility of terminating the network connectivity in a radical manner, that is, by simply disconnecting the home cyber systems from their network cables, and in the case of a home wireless network, disconnect the home wireless router. This solution, though radical, may minimize the extent of the damage to the home computer(s), since the speed of the attacking code vector may exceed the speed of the countermeasures generated by the home network owner. As in the case of a network enterprise, the home user must have an in-depth knowledge of the behavior of his home network. This knowledge will become the critical gage in determining when something unusual and potentially harmful may affect the home network.

The knowledge and expertise of conducting forensic analysis on raw network data packets is the most effective means of understanding the attacking strategy of the adversary, and also the best tool for establishing our own defense, knowing exactly the modus operandi of the attacking strategy. Of course this knowledge is reserve for those qualified cyber SMEs dedicated to protect and defend a network enclave. Because of the reduced number of these qualified SMEs, and the lack of vision on both industry and government on increasing and retaining the contingent of these SMEs, the vendors have appealed to the large enterprises with a band-aid solution marketed under the moniker of Deep Packet Inspection (DPI). The vendors, capitalizing on the fixation and implicit trust of large enterprises on automated systems, are offering network appliances capable of performing analysis of network data packets on real-time, and inspect the packets from layer 2 through layer 7.[89]

While is true that most current DPI appliances are capable of offering this real-time analysis of the network packets traversing the boundaries of a network enterprise, the unchecked enthusiasm of some technical writers lead them to hyperbolize the impact of DPI by characterizing them as "the next-generation technology"[90] that came to fill the void created by intrusion detection systems, that were unable to inspect every packet.

The statement of the writer hyperbolizing the benefits of DPI is incorrect at two different levels. First, DPI is not the next generation technology, but simple the automation of a process already used by qualified cyber SMEs decades earlier than the appearance of DPI, specifically, from the time when tcpdump[91] became available as a packet analyzer. Any packet analyzer allows the qualified cyber SMEs to conduct comprehensive research of the network traffic monitored and/or captured with the packet analyzer. Second, the only difference between DPI and a qualified cyber SME is the amount of network traffic that can be monitored on a given amount of time. While a DPI is substantially faster than a cyber SME, the DPI can never match the deductive and analytical powers of a cyber SME.

---

[89] The Open Systems Interconnection (OSI) abstract model consisting of seven layers forming the framework for networking architecture standardized under the aegis of the International Organization for Standardization (ISO)

[90] http://www.techrepublic.com/blog/networking/deep-packet-inspection-what-you-need-to-know/609

[91] Originally written in 1987, and widely used since the 1990s.

The impact of conducting analysis on raw network data, during and after a cyber attack, is invaluable. During the attack the defending cyber SMEs can learn and even anticipate the strategic phases of the attack. This capability is critical in regaining the initiative, which is the prelude to gain advantage over the adversary. After the cyber attack, the analysis of the network traffic can be use to modify, enhance and harden the cyber defenses of the network enclave, and the captured adversary network traffic can even be implemented into replay exercises, to test and improve the revised cyber defense plan.

In summary, the most important concept and policy in any cyber defense plan is to reassign levels of operational authority, based on the principles outlined during the discussion of CIRP and LCDS. The synergy and coordination of all boundary system administrators and the LCDS during a cyber attack is the most crucial foundation to aspire for network and cyber survivability.

Chapter 14. The ultimate equalizer

The knowledge and skilful practice of the cyber principles act as the ultimate equalizer in our digital age. No natural resources, financial strength, or powerful military force can stand against the almost unassailable power of the cyber spectrum. However, cyber is the product of the digital age, and there is a more powerful threat that can annihilate cyber prowess. The Achilles heel of the electronic age, and by extension, of the cyber age, is the electronic infrastructure.

The very essence of the electronic age is extremely fragile since it depends on the availability and accessibility of the electric power grid feeding our insatiable appetite for electricity. But even though we can design and implement powerful cyber defenses, there is no defense against a regional, or even worse, a global blackout. While man-induced blackouts are a weapon[92] deployed either as an act or war or terrorism, there is a far worse type of blackout; the one produced by the powerful electromagnetic pulse (EMP) effect generated by a solar storm, against which there is no possible defense, since the power level of such EMP is exponentially higher than any feasible human defense.

---

[92] A quote from an official newspaper of the PLA speaks of a surprise attack on US critical information systems by means of electromagnetic pulse weapons, stating that "when a country grows increasingly powerful economically and technologically . . . it will become increasingly dependent on modern information systems . . . The United States is more vulnerable to attacks than any other country in the world" (Zhang Shouqi and Sun Xuegui, Jiefangjun Bao, 14 May 1996). Quoted by Dr. Peter V. Pry, EMP Commission Staff, before the US Senate Subcommittee on Terrorism, Technology and Homeland Security, March 8, 2005

The intense energy field generated by an EMP event can instantly overload and disrupt electrical circuits, and modern microcircuits are especially vulnerable to the enormous power surges generated by EMPs. While human-induced EMP events can be produced at a large scale using a single nuclear explosion[93], and on a lesser, non-nuclear scale using battery devices or chemical explosives[94], these EMP events, introduced by either nation states and/or terrorist entities as part of cyber conflict strategies, cannot reach the critical level of global destruction of the electronic age infrastructure.

A HEMP event creates gamma-radiation that interacts with the atmosphere to create an intense electromagnetic energy field capable of destroying electronic circuitry with a level of intensity exponentially higher[95] than a lightning strike. Any nation with a high degree of dependence on computer networks and the power grid will suffer a catastrophic and prolonged disruption of the critical infrastructure when confronted with an EMP event. On July 2004, when the "9/11 Commission Report" was made public, the US Congress was warned that as a nation we stand virtually unprotected against the effects of an EMP event "that could damage or destroy civilian and military critical electronic

---

[93] As in the case of High-Altitude Electromagnetic Pulse (HEMP), generating an electromagnetic energy field produced in the atmosphere by the power and radiation of a nuclear explosion, extremely damaging to electronic equipment over a very large area

[94] Such as in the case of High-Power Microwave (HPM), the electromagnetic pulse produced with special equipment that transforms power from batteries, or from a chemical explosion, into intense microwaves that are very destructive to electronics within a smaller area.

[95] HEMP effects became extensively known in 1962 during a high-altitude nuclear test (code named "Starfish Prime") over the Pacific, when radio stations and electronic equipment were disrupted 800 miles away throughout Hawaii

infrastructures, triggering catastrophic consequences that could cause the permanent collapse of our society."[96]

It's highly ironic that older electrical components (vacuum tubes) are generally more tolerant to electromagnetic pulse, but our constant search for miniaturization of electronic components renders modern circuitries increasingly more vulnerable to electromagnetic interference. Thus, countries with infrastructures built on older technology may have a better probability of surviving the disabling effects of HEMP or HPM than countries relying on modern electronic technology[97].

The devastating and persistent destructive power of an EMP event emanates from the peculiar characteristics of such an event. An EMP event adversely affects all electronic systems in general; it's effective in all weather conditions and in all terrains and area sizes, and produces permanent destructive effects in all electronic circuits. Hardening of electronic equipment is a costly predicament and offers limited mitigating results. Some school of thought consider that man-induced EMP events can be launched with impunity since they result in minimum human casualties (if any), thus minimizing potential political and military repercussions[98]. However, a man-induced event targeted at an ICS may result in a heavy count of casualties.

The possibility of a devastating solar storm creating a destructive EMP event is not a scientific hypothesis, but it's based on historical scientific evidence. The events leading to the most severe solar storm in recorded history happened in 1859, and together they

---

[96] Electromagnetic Pulse (EMP) paper presented by Joe St Sauver, Ph.D., at the Infragard, Eugene, Oregon, Eugene Public Safety Center, April 30th, 2009

[97] Lowell Wood statement before the House Research and Development Subcommittee, hearing on.EMP Threats to the U.S. Military and Civilian Infrastructure, Oct. 7, 1999.

[98] "Electromagnetic Pulse Threats in 2010", by Colin R. Miller, Major, USAF, Center for Strategy and Technology Air War College, Air University, November 2005

caused the most powerful disruption on the Earth's ionosphere, as attested by NASA scientist studying the phenomenon, because its significance is not only as a historical landmark, but also as a 150 years old event that can happen again.[99]

The Sun radiates energy at an average of approximately 400 billion trillion kilowatts. On a schedule not yet understood by the scientific community, the Sun releases tremendous outbursts of energy in the form of solar flares, namely, coronal mass ejections (CME) formed as the result of an explosive release of a massive cloud of magnetically charged plasma. The effects of these solar flares on the Earth ionosphere become visible and are popularly known as Northern Lights. This unusual display during the super solar storm on 1859 became visible not only on the northern hemisphere, but as far south as Rome and Hawaii. The display of aurorae was observed around the world, and since customarily they are only visible to people in the polar region, they were misunderstood elsewhere. The aurorae were so bright that gold miners in the Rocky Mountains misidentified the display with the dawn, and set up to eat breakfast at 1 AM, thinking it was a cloudy day.[100]

The image below shows the Solar and Heliospheric Observatory (SOHO) image of a recent CME heading toward Earth on October 22, 2003. The CME arrived on Earth on October 24, and caused a geomagnetic storm of less severity than the super solar storm of 1859. The fact that the effects of this 1859 super solar storm did not reach catastrophic effects was due to the fact that 1859's society was living through the very early steps of the electronic age, and their dependency on electricity was very low, thus limiting the negative impact to the nascent telegraph system, that became disabled across Europe and North America.

---

[99] http://science.nasa.gov/science-news/science-at-nasa/2003/23oct_superstorm/

[100] http://www.scientificamerican.com/article.cfm?id=bracing-for-a-solar-superstorm

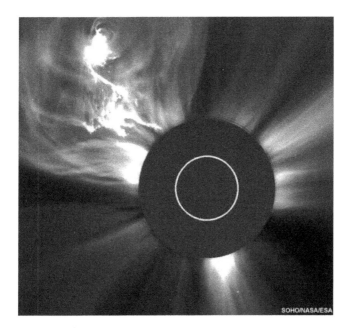

SOHO/NASA/ESA

More recently, the effects of solar storms have escalated the amount of damages to our electronic infrastructure. The 1989 solar storm disrupted the Hydro-Quebec power grid in Canada for over nine hours, with damages and losses in revenue estimated in the hundreds of millions of dollars. In 1994 another solar storm was responsible for major malfunctions on two communications satellites, and disrupted the operations of newspapers, network televisions and radio services throughout Canada. Other recent solar CME events have affected cell phone services, TV signals, GPS systems, and electrical power grids.

As we can see from this brief historical review of the recorded solar storms targeting the Earth with EMP events, the next CME reaching equal of higher levels than the one in 1859 will set our global electronic village back to the pre-industrial age. While is understandable and judicious that we place a lot of interest in keeping a vigilant eye on a man-induced premeditated EMP event, it is shortsighted to see this eventuality as the only threat. While any nation with the required industrial and military means can generate an EMP event targeted at another nation perceived as a threat, this type of event is regionally limited, and will affect only the targeted region. The next big solar storm will affect the entire planet, and if we are not prepared as a global village to face this inevitable next global EMP event, all nations will be plunged into the pre-industrial age in a fraction of a second, and we'll remain there for a very long time. The entire electronic infrastructure on planet Earth will disappear in an instant, and unless we have a contingency plan when this occurs, the catastrophic effects will become almost fathomless.

This line of thinking is not a gratuitous speculation indulging in baseless fears, but rather a sober assessment of the historical data collected by the scientific community. The next large solar storm, unpredictable but inevitable, will effectively erase the cyber environment we have built on the electronic infrastructure sustaining our modern society. All we can do is to design a contingency plan on how to continue our existence without the cyber environment on which we have become so utterly dependent. Unless this book is printed before the next large solar storm, there will be no record of it because at the time of its writing it exists only in the cyber spectrum segment generated in my computer. Likewise, all financial, medical, industrial, scientific, military, artistic and a vast number of our current society records created and stored in the cyber spectrum will disappear in an instant so brief[101] and fast than we will not know it happens until the aftermath of the event itself.

The fallacious belief that the Internet is an eternal and indestructible cyber entity is fed by the ignorance of the masses using the services offered by the Internet[102]. One of the many services offered by the Internet is the World Wide Web (WWW), which should not be confused or considered synonymous with

---

[101] The rise time of an EMP event (the time it takes for the pulse to reach maximum amplitude) is less than a few thousandths of a second, and thus too fast for any protective circuitry to react in time. Furthermore, this is only an academic consideration, since the alleged protective circuitry itself will also fail due to the impact of the same EMP event.

[102] For a brief technical definition of the term "Internet" see chapter 3.

the Internet. During the early 1990s Tim Berners-Lee, a scientist at CERN (Conseil Européen pour la Recherche Nucléaire)[103] invented and deployed the WWW in 1992, as a facilitator for the information exchange among physicist around the world. The WWW is based on the Hypertext Transfer Protocol (HTTP), the underlying and foundational network protocol designed for the distribution and collaboration of hypermedia using this hypertext model.

When in February 2011 the bookstore chain Borders filed for bankruptcy, and announced the closing of all their stores, a plethora of printed and broadcasted media commentaries were disseminated with pseudo and shallow analysis about the role "of the Internet" in the demise of Borders, and a local reporter labeled the Internet as "the monster that's creating the problem"[104]. This illustrates the popular misconception of equating the Internet with the WWW, and underlines the fallacy of assuming that the digital age and the convenience of obtaining all our information via the WWW will remain forever. The demise of Borders does not represent the death of printed literary products. When the next CME occurs and our entire digital age disappears with the demise of the Internet infrastructure supporting the WWW, the printed page will remain the foundation of our knowledge repository.

How do we continue functioning after the cyber environment disappears, obliterated by a massive solar storm that destroys our electronic infrastructure at a global level, affecting all nations instantly and indiscriminately? That is the big next question for the planet Earths inhabitants, so accustomed to the support and convenience currently offered by the cyber environment sustaining our modern life styles. Careful monitoring of the intentions of other nations seeking to launch a man-made EMP event will allow us to predict when such attack is unleashed, and we certainly will be able to recover from it. But against a massive CME affecting our entire planet, even when we may have a window of less than 24

---

[103] European Council for Nuclear Research, in Switzerland.
[104] Las Vegas Sun, "The man who killed Borders", Sunday, July 31, 2011, Commentary, Page 7

hours to prepare ourselves for the EMP event, there is absolutely nothing we can do to either avoid or mitigate the catastrophic event. Humans will survive the EMP event, but our global electronic infrastructure will not. Without the modern umbilical cord attached to an obliterated electronic infrastructure, how can we sustain the very fabric of modern life, with its high dependence on the trucking and transportation industry supporting commerce and food distribution, fuel refineries, and without the foundation provided by the information and communication, and military capabilities?

Excursus: Cyber isolation by self-determination

During the writing of this chapter an unprecedented cyber event took place in the world scene: a nation decided to exercise the right to self-determination and severed all cyber contacts with the rest of the world. While is true that such action is perfectly feasible from a technical point of view, prior to January 27, 2011 nobody really considered that option, and much less, exercised that right. Why? Perhaps because such an action was thought to be counterintuitive, and contrary to the trend and the inertia of cyber connectivity, that has become the goal of every nation and individual since the advent of the World Wide Web in the early 80's. Popular wisdom dictated that everyone wanted to established cyber connectivity with the rest of the world, and no one entertained the thought of seeking cyber isolation as a voluntary action of self-determination.

Ever since the first connection between two computers across the continental USA in 1969, the launching of ARPANET, and consequently the design and implementation of TCP/IP in the late 1970s, the interconnection of computing systems and cyber components around the globe has become a widespread cyber phenomenon. And yet, inevitably, all this exponential growth will come to an abrupt end with the next mayor EMP produced by the imminent CME dictated by the solar cycle. Do we have a feasible contingency plan to continue operations in our interconnected global village? Is there a feasible contingency plan?

Chapter 15. Hybris and sophrosyne

Like in any other form of conflict, an excessive propensity to pride, self-confidence and arrogance may lead to defeat. In a cyber conflict is extremely important to maintain our cyber technical prowess under control, so that the technical knowledge may not lead us to defeat by underestimating the cyber adversary, and failing to keep the proper balance between assertiveness in victory and prudence.

The concepts of ὕβρις (hybris)[105] and its antithesis σωφροσύνη (sophrosyne) are very relevant into a book dedicated to cyber conflicts, where two parties are involved in a confrontation in cyberspace. The former characterizes the attitude of being insolent and arrogant due to an excessive self-confidence that diminishes the opponent, while the latter characterizes the demeanor of the one that engages in the conflict with an attitude regulated by prudence and self-control, with a degree of moderation balancing knowledge and assertiveness.

Since the recipe for victory in a cyber conflict is knowledge and experience, there is always the risk of letting one's knowledge becoming the seminal root for arrogance, leading to the always fatal mistake of underestimating the adversary. The famous play

---

[105] This author prefers the transliteration "hybris" over the popularized "hubris", and for a very simple reason; consistency. Both Greek spellings for these antithetical terms contained the letter upsilon, and upsilon is traditionally transliterated as "y" in the Western alphabet, thus the transliteration of Κύριος as Kyrios (Lord), and gynecology, derived from ancient Greek gyne, γυνή (modern Greek gynaika, γυναίκα) for the medical study of women. Since σωφροσύνη is transliterated of sophrosyne (using the letter "y"), there is no logical reason to transliterate ὕβρις as hubris.

produced by the tragedian Sophocles around 430 B.C. presents the dangers of hybris in his acclaimed work "Oedipus Rex", the story of Oedipus, who, driven by hybris, killed his (at the time unknown) father over a trivial dispute on who has the right of way at an intersection, and later defeated the ruthless Sphinx guarding the entrance to Thebes, and, while still fueled by hybris, he accepts the crown as king of Thebes and marries his own (at the time unknown) mother. The tragic ending of this story sees Oedipus learning the horrible truth about his father and mother, and blinding himself. All this tragic events in Oedipus' life were cause by his free will in exercising hybris.

During a cyber conflict we are always confronted by the nonlinearity of the seemingly chaotic, complex and confusing flux of events. In order to discern through the complexity and nonlinearity of these events one must keep a cool and level head, always considering that the actions of the adversary may be orchestrated by a deception factor. Hybris may cause us to act with arrogance and over self-confidence, and while we must be assertive in our actions, prudence and balance must prevail in our decision making process. Frequently our hybris sets us up for defeat even before facing a cyber conflict, as in the infamous case when we designed new systems with known cyber security flaws, and we deploy such flawed systems assuming that our potential adversaries are not skillful and resourceful enough to design and implement techniques allowing them to profit from our hybristic attitude. One of the most recent examples comes from the UAV data intercepted by militants in Iraq and Afghanistan.[106] We created a surveillance system to allow us the gathering of intelligence data on the actions and movements of the adversary,

---

[106] On December 17, 2009, the Wall Street Journal online : "The potential drone vulnerability lies in an unencrypted downlink between the unmanned craft and ground control. The U.S. government has known about the flaw since the U.S. campaign in Bosnia in the 1990s, current and former officials said. But the Pentagon assumed local adversaries wouldn't know how to exploit it, the officials said." The Shiite militants used the COTS software SkyGrabber, available for $26.00 on the Internet.

but our hybristic miscalculations have allowed the adversary to profit from the same data, thus revealing the areas we keep under surveillance. This case is a perfect example of the premise of this book: cyber as the ultimate equalizer. A powerful nation creates a sophisticated surveillance system, which is deployed without the proper cyber security requirements, because we deemed the adversary as lacking the technical sophistication required to intercept the surveillance data. The adversary, even though significantly less powerful and sophisticated, used the available cyber tools in the open market to intercept the surveillance data, and to profit from the tactical and logistical value of the collected information. The hybristic miscalculation of the powerful nation has become a tactical and logistical advantage for the less powerful adversary, by using cyber as the equalizing factor.

Sophrosyne, the antithesis to hybris, should not be misunderstood as inactivity or indolence. The quality of applying prudent and balanced calculations to our assessment of the adversary's capabilities should also apply the principle of cyber as the ultimate equalizer. A comprehensive knowledge of the cyber ROE is an invaluable asset when confronting a cyber adversary, and mounting countermeasures to cyber attacks. Lacking this degree of knowledge on cyber ROE can have a paralyzing effect. A few years ago this author was invited to participate in the writing of a series of tactics, techniques and procedures designed to enhance our cyber defense posture. During one of the writing sessions a participant shared his frustration for having to endure persistent cyber attacks from an unknown adversary that was decimating their cyber defenses. After providing details on such persistent attacks it became perfectly clear to this author that the victims were not using DNS[107] resolution to elucidate the identity and location of the attacking systems. When this author advised the participant to use DNS resolution data to mount a cyber defense countermeasure, the participant was shocked at such suggestion,

---

[107] The Domain Name System (DNS) facilitates the assigning of domain names and mapping them to corresponding IP addresses, through a hierarchical naming system built on a distributed database for cyber systems connected to the Internet.

since he was convinced that utilizing DNS resolution data was an illegal action that his organization was not willing to undertake.

This case illustrates how an individual, and the organization depending on him, was paralyzed by the misunderstanding of the nature of DNS and the ROE regulating the use of the DNS data. DNS is a public domain database, searchable by anyone, and the identification of any particular Internet resource registered in this public domain database does not constitute an illegal action. This lack of understanding on the proper and authorized use of the DNS data, and the corresponding cyber ROE, had reduced this organization to suffer a persistent cyber attack without initiating any countermeasures to protect itself.

Sophrosyne brings into proper balance the fitting response to any cyber activity, and within such balance cyber knowledge and expertise are kept in a state of equilibrium with the permissible cyber ROE. Sophrosyne is equally important when taking a cyber initiative or responding to an adversarial cyber initiative. However, there is a reduced advantage in the latter scenario. Sophrosyne works best when we have had the opportunity to study and assess the adversary in order to avoid underestimating our opponent. When we are placed in the predicament of having to react to the adversary's initiative, then we may not have the opportunity to profit from an assessment of the opponent's capabilities, since we'll be primarily occupied in defending ourselves against the adversarial initiatives.

The corollary to this brief discourse regarding hybris and sophrosyne is that is always more profitable and advantageous in a state of cyber conflict to take the initiative whenever feasible, since the initiative allow us to dictate the battle rhythm, after proper consideration to the dangers linked with hybris and the advantages associated with sophrosyne.

When it comes to striking a balance between our expectations and the feasibility of victory in a cyber conflict we need to factor in the concept of mediocrity. In the context of a cyber conflict, victory can never be achievable by simply "doing what it is required;

no more, no less". There are too many organizations based on this mediocre standard, trying to get by simply by doing just the minimum required. Required by whom? Defined by whom? What is indeed the "minimum" required to achieve victory in a cyber conflict? This author participated in a series of cyber security assessment studies requested by an organization seeking "a bill of health" on their cyber security posture. After completing the study, based on the dissection of the set of vulnerabilities and weaknesses assessed by a team of cyber SMEs, and corroborated by documented cases of cyber intrusions and a history of remediation activities, a final report was delivered to the requesting organization, with a somber warning message, stating a suggested list of the new and additional cyber security measures needed to be implemented in order to achieve a feasible and acceptable cyber security posture. The requesting organization received the final report, but declined to implement the recommendations. What was the rational for their declining? "We are already complying with the minimum cyber security measures required by the current standards."

How did this organization arrive at the current state of cyber insecurity? The state of cyber insecurity was achieved by adhering to the comfortable position of doing no more and no less than the minimum required by the current standards. And who designed those standards? The minimum standards were designed by a group of individuals with a marginal understanding of what constitute a cyber conflict, in agreement with a group of advocates seeking to deflect the responsibility of cyber defense to a series of automated tools, instead of assigning such task to knowledgeable and experienced cyber SMEs. This organization was confident that a group of remotely monitored sensors would provide them with a cyber defense strong enough to keep their local network safe. Ultimately, this position relies on the fallacy that remote monitoring provides enough cyber security. What is the fatal weakness in this position? To ignore the rule of thumb in cyber security that states that only those with first hand knowledge on the local network can successfully defend it.

It is hybristic to assume that our network is safe because we are not aware of any problems. We must consider that the nonlinear character of cyber activity demands we keep constant monitoring on our network activity, because only the knowledge of anomalies in the behavior of our network will provides us with the indicators pointing to an emerging cyber threat. And being that the measure of normalcy is a function of the complexity of our cyber interaction with the complex Internet environment, a true and effective cyber monitoring should not be solely entrusted to automated cyber systems, because the very measure of normalcy in a given network is never a static function, but a highly dynamic one. Automated monitoring systems can only measure a static normalcy baseline, but a dynamic normalcy can only be monitored and determined by cyber SMEs, equipped with technical knowledge and analytical skills, and the knowledge of the behavior of the network enclave entrusted to their protection.

The amazing awakening that our society experienced with the advent of the Chaos Theory represents a tremendous advantage, empowering us with the appropriate tools to perform network normalcy behavior analysis, since chaos theory is essentially the acquired cognition about finding the underlying order in apparently random and complex data systems. When meteorologist Edward Lorenz, the pioneer in chaos theory, began in 1960 his research on weather prediction with the help of a computer designed to create weather modeling, he unveiled the amazing principles underlying the interaction of elements within complex dynamic systems (CDS), characterized by a myriad of elements in motion, generating numerous possibilities in the outcome. When Lorenz set in motion a previously recorded sequence, but modified it by starting it at a middle point, the previous sequence evolved differently, creating a pattern completely different than the recorded original pattern.

The new pattern was caused by the mathematical variables provided on the second run of the recorded sequence. The original variables were stored with six decimal places in the computer memory, but Lorenz run the second sequence with only three decimals, expecting to expedite the process and obtain a reasonably close approximation to the original sequence. This is the moment when chaos theory was born. An apparently miniscule change may have unpredictable results, and the corresponding effects may not be commensurable with the magnitude of the changes.[108] This phenomenon is known as sensitive dependence on initial conditions in chaos theory parlance. Chaos theory discloses the unpredictability of CDS and displays them in patterns. Thus our traditional linear way of thinking was forever shattered with the advent of this new unveiled reality of our universe, of which chaos theory offers us a glimpse.

Prior to the arrival of the quantum mechanics revolution we were used to believe there was a linear and sequential relationship between cause and effect, and even Freud based his psychoanalysis premises on this simple causation, teaching that current anomalous states in the human mind were the results of a psychological experience in one's past. Quantum mechanic's Uncertainty Principle states that absolute accuracy is not attainable, and since the initial situation of a CDS can not be accurately determined, neither can the evolution of a CDS be accurately predicted.

It is sophrosynistic to seek a factual assessment of the strengths and weaknesses in our network, and to undertake a thorough assessment of the existing risks, while categorizing them as the ones we must not take under any circumstances, and assume an acceptable loss on those risks we cannot control. No network is invulnerable, but every network is defendable. We simply have to apply prudence, wisdom, soberness in our assessment, and avoid hybris and indolence. The only acceptable means of knowing our network is by discovering and categorizing our network weaknesses. Once we decide what our defensive priorities are, then we must research the appropriate remediation plan and apply the available solutions. This is proactive defense, the result of a sophrosynistic resolve to stand before the adversaries, and not succumb to their attacks.

---

[108] This has been dubbed as "The Butterfly Effect"

Network performance and network behavior is a complex entity, but can be codified by measuring critical variables, resulting in network performance and network behavior baselines. A sustained and dedicated monitoring of these baselines is a valuable tool not only on measuring the productive health of a network, but also a valuable diagnostic analytical tool for cyber defense. Once these baselines become a pattern depicting a normalcy measurement, they will allow us to determine anomalous network behavior that may point to the presence of a cyber attack. Let's use DNS behavior patterns as an example.

If we have had the foresight of creating a baseline on the behavior and performance of our DNS server(s), then we can monitor this baseline and determine when unexplainable anomalies occurred. Anomalies in themselves do not directly correlate with a threat, but anomalies in DNS behavior do indicate the possibility of an exploit. We also have to have a mature understanding of our DNS behavior in order to differentiate changes from anomalies. A change is an explained alteration in the behavior of our DNS baseline caused by an authorized change in our network. An anomaly is an unexplained alteration in the DNS baseline. An unexplained alteration in the volume and frequency of our DNS query stats, or an unexplained variation in the number of DNS query failures on our external DNS servers, may indicate the presence of an exploit. A large number of failed A record queries for domains where our DNS servers are not authoritative, especially when they are configured not to perform recursive queries, may be an indicator of a large number of compromised internal nodes attempting to connect to an adversarial set of C2 nodes.

Here is where the demarcation between reactive cyber defense and proactive cyber defense is established. The majority of those organization that rely solely on automated systems for their intrusion detection system (IDS) program will eventually find that the answer to destructive intrusions aimed at them was clearly outlined in their DNS logs, but such valuable information was either ignored because of lacking analytical skills, or because of plain ignorance. How many organizations really exercise and practice proactive cyber defense? Please don't answer this rhetorical question, since we already know the gloomy answer. Those very few organizations with the foresight to implement and sustain a proactive cyber defense program will have dedicated and expert cyber SMEs performing monitoring and analysis on their DNS logs, and they will be able not only to detect cyber exploits in progress, but in many cases they will be able to implement the necessary countermeasures to defeat an incipient exploit.

> No network is invulnerable, but every network is defendable. We simply have to apply prudence, wisdom, soberness in our assessment, and avoid hybris and indolence.

There are so many organizations grouped in the former category of placing exclusive dependence on reactive defense that they think proactive defense is an unattainable ideal, and they believe that a preventive cyber defense is a Hollywood fantasy. These were exactly the words of a pseudo-cyber individual who used the words "a Hollywood fabrication" to dismiss a report presented by this author during a cyber conference, outlining the use of DNS records in order to detect and nullify a DDoS. These types of individuals and organizations are so accustomed to experience defeat that they actually think victory in a cyber conflict can only happen in the movies.

Sometimes hybris may be the result of ignorance, and sometimes the result of an overly optimistic perception of our knowledge. The road to cyber victory is often found on the foundation of a realistic and pragmatic assessment of our empirical technical knowledge, and a sober assessment of our cyber adversary. We must strive in avoiding the extremes of underestimating our nemesis, or adopting an a priori defeatist attitude because we perceive our adversary as possessing superior cyber knowledge beyond our reach. Cyber is the equalizer because cyber knowledge and cyber skills are available to everyone. Cyber excellence is not bestowed by an official title. Cyber excellence is the result of dedication to learn and work hard to achieve a level of proficiency that exceeds that of our adversary.

Chapter 16. Exascale and Quantum Computing

In closing our discussion on cyber as the equalizer we come to the point where we look forward into the development of cyber technologies. The current demand for advances in High Performance Computing (HPC) technology is vectoring our efforts through two different paths; the traditional progression within the scope of the original binary architecture, and the new computing paradigm based on the discovering of the behavior exhibited by sub atomic particles, allowing us to redirect our efforts into the realm of quantum computing (QC). In the traditional binary architecture our nation has already lost the leadership in HPC at the petascale level. On the quantum computing level, however, there is no current leader on this race, but the runners are numerous and eager. Attempts to recapture the leadership at the petascale level do not represent a feasible solution, since this level of HPC is no longer able to satisfy our scientific and technological demands, and consequently we need to focus our efforts on achieving HPC performance at the exascale level. In order to maintain scientific, technological and military global superiority, it is highly desirable for the U.S. to establish HPC leadership both at the exascale and the QC levels.

There are great economic and strategic benefits in achieving exascale HPC, because these two factors pave the way to technological advancement and military superiority, and the U.S. is not the only runner on this race, which includes the current two leaders on petascale, Japan and China, plus the European Union, Russia, and India.[109] The ascendance of Asian countries

---

[109] Chris Nutall, "Supercomputing's exascale arms race," October 15, 2011, http://blogs.ft.com/fttechhub/2011/10/the-exascale-supercomputing-

to the top two leadership positions in HPC is not an anomaly, but rather the demonstration on how the epicenter of computing excellence has migrated from North America to Asia.

Achieving HPC at the exascale level is intimately connected with the successful design, testing, and production of high energy devices, more efficient and powerful engines, optimized and stealthier flying platforms, revolutionary materials, superior weaponry, and national, economic and environmental security.[110] This overall enhanced strategic progress is attainable via the use of powerful and versatile modeling and simulation (mod&sim) platforms operating at the superior level offered by exascale HPC with a performance rate of 10^{18} Eflops/sec. The current operational petascale systems (10^{15} Pflpos/sec) are insufficient for the computational demands presented by the mod&sim requirements necessary to achieve the enhanced strategic technological progress essential for undisputed superiority. If successful, the attempt initiated by Oak Ridge National Laboratory (ORNL) to recapture HPC leadership at the petascale level may become an ephemeral and pyrrhic victory;[111] strategically is more advantageous to redirect the efforts into achieving leadership at the exascale level.

In addition to the quest for superiority in HPC via the traditional binary path of exascale, we must also seek superiority in achieving operational leadership on the sub-atomic path of QC, allowing us to surpass the limitations imposed by the binary architecture, where we remain limited to only two states and a single choice between these two states. On QC we are allowed to transcend that limitation by being ushered into a cyber realm where the computational universe is limited only by the number of qubits we can employ,

---

arms-race/?catid=677&SID=google#axzz1azNs4raE

[110] Rick Stevens and Andy White, *A decadal DOE plan for providing exascale applications and technologies for DOE mission needs.*

[111] Dawn Levy, "Oak Ridge National Laboratory Awards Contract to Cray for 'Titan' Supercomputer," October 11, 2011, http://www.olcf.ornl. gov/2011/10/11/oak-ridge-national-laboratory-awards-contract-to-cray-for-%E2%80%98titan%E2%80%99-supercomputer/.

and we become empowered to operate simultaneously within all the multiple choices generated by the number of qubits employed by a given QC system. We are already benefiting from a reduced subset of quantum applications, especially in the critical field of cryptography, bur we are not alone in these fields.

The importance of using petascale HPC in scientific research and technological development is quite evident in the June 2011 version of the official list of the top 500 HPC systems, where Japan, China, UK, Germany, France, and Russia are the nations with double-digit listings. Of all the 500 listed systems, 30% declared a cryptic unspecified application area, while 15% is declared as dedicated to research, and 4% dedicated to defense. It is not unreasonable to extrapolate that 30% into classified research and defense projects.[112] In our global and compulsory quest for fossil fuel the Chinese Feoso case attests to the strategic advantage of petascale HPC by reducing calculation time from 6 months to only 16 hours.[113] On the other hand, the modeling of the structure of an entire airplane interacting with atmospheric variables and sound waves under hypersonic conditions during mission maneuvers is an unattainable goal with the current petascale HPC systems. The solution requires an order of magnitude increase in computational capabilities as the ones provided by exascale HPC systems.[114]

The gains already achieved by using mod&sim at the petascale level allow us to project the enormous advantages of transitioning into the exascale level to overcome the current limitations present at the petascale level. The computation of the RCS[115] of an entire modern fighter platform, given a fixed incident angle and a 1 GHz

[112] http://www.top500.org/
[113] TradeKool: Global Business News, http://biznews.tradekool.com/13909/1/Race_is_on_for_new_generation_of_supercomputer.html, August 20, 2011
[114] DARPA IPTO, AFRL contract number FA8650-07-C-7724, "Exascale Computing Study: Technology Challenges in Achieving Exascale Systems, September 28, 2008.
[115] Radar cross section, a measure of radar detection for a given object

radar signal, remains unattainable with the present petascale systems. The introduction of exascale HPC can translate this issue into an attainable goal.[116] The technological advances achieved via the research into nanotechnology feed the progress into HPC as well, as demonstrated by the computing architecture of the Japanese HPC system K, currently number 1, delivering 10.51 Pflops and using 45nm Sparc64 processors, as an example of nanotechnology development assisting petascale HPC.[117] Petascale has already provided engineering solutions critical to national defense, as in the case of simulation of a complete gas turbine chamber that reproduced the "combustion instability" phenomenon, responsible for catastrophic engine failures in helicopters, rocket, and aircraft turbines. The official report[118] of the Japanese HPC manufacturer states that the K system applications, once fully operational on the second half of 2012, will include nanotechnology, disaster prevention, aerospace, life sciences, nuclear power, and astronomy and astrophysics, among others. Yet, many other challenges in the scientific and technological realms remain unapproachable. Therefore, the transition into exascale HPC is critical to national security in scientific, technological, and military endeavors, such as in aiding a design framework for producing optimized flying platforms and the corresponding advanced propulsion systems.[119]

The path to exascale is not a straight road on a two-dimensional plane, but rather a multiplicity of avenues on multi-dimensional planes. Moore's Law is no longer relevant in the exascale realm, primarily because of the power factor. Extrapolation of the power requirement for scaling up current systems to perform at the exascale level would surpass the gigawatt level, and that it's not a

---

[116] Ibid

[117] http://www.zdnet.co.uk/news/emerging-tech/2011/06/21/inside-japans-top5000-k-computer-40093162/

[118] Fujitsu Report. "Approach to Application Centric Petascale Computing", Motoi Okuda, Fujitsu Ltd., 16 Nov 2010"

[119] U.S. DoE, Office of Science, "Summary Report of the Advanced Scientific Computing Advisory Committee (ASCAC) Subcommittee, Fall 2010."

feasible solution.[120] Additionally, we need to engage in fundamental research and discovery of new types of processors, operating systems, networking paradigms, and memory management schemes. This latter problem is clearly illustrated by the fact that we have achieved improvements in CPU cycle time, but without the corresponding improvements in memory access time.[121]

The concept of scaling is also becoming rapidly obsolete, as power consumption and heat generation reach critical levels. Consequently, and in order to achieve the required improvements paving the way to the exascale level, we must find solutions to reach a 100 times energy consumption reduction factor, new and enhanced parallel programming paradigms, radical innovations in networking technologies, and new interconnects designs. Pursuing the scaling of progressively larger CPU-powered systems is no longer the answer, due to the energy consumption factor. Thus, one of the avenues with a promising outlook is the design of the heterogeneous model, where CPU and GPU coexist, combining the corresponding advantages of the CPU optimization for linear tasks, and the GPU optimization for parallel computing, while offering a more energy efficient operation. Combining these two types of processors will certainly pave the way to exascale development.[122]

The emerging of the novel OpenACC programming paradigm will facilitate the adoption of the heterogeneous CPU-GPU model, with the OpenACC directives enabling the compiler to facilitate the mapping of computations to the accelerator. This new open parallel programming standard was unveiled during the SC11 conference.[123] A similar hybrid approach is being explored in the

---

[120] The Opportunities and Challenges of Exascale Computing. Sumnmary Report of the Advanced Scientific Computing Advisory Committee (ASCAC) Subcommittee, Fall 2010, p. 2, 49

[121] Ibid, p. 50

[122] http://blogs.nvidia.com/2011/11/exascale-an-innovator%E2%80%99s-dilemma/

[123] The International Conference for High Performance Computing, Networking, Storage and Analysis, convened in Seattle, November

ARM processor, capable of operating with less power than the traditional x86 processor, though lacking the performance required for HPC operations. However, combining ARM CPU processors with the high performance of GPUs we can achieve the goal of high energy efficiency without performance compromises.[124] Additionally, run-time errors will present another formidable challenge. Currently, we are coping with 0.5 million processing elements in current systems, under manageable conditions, but the escalation of run-time errors into the exascale realm, with an expected 1 billion processing elements, will increase the frequency of run-time errors to a level that may become extremely difficult to manage. However, these challenges must be overcome, because of the phenomenal gains awaiting us in the exascale realm. The ASCAC report[125] offers a detailed description of the exascale computing benefits for the aerospace and propulsion industries, astrophysics, biology, climate, materials, fusion energy, nuclear energy, and national security.

The traditional binary computational paradigm, even at the current HPC petascale level, is limited to a single choice when performing calculations, by opting for either a 0 state (off) or a 1 state (on). In QC a qubit offers three choices; a 0, a 1, or both simultaneously, because it's based on the laws of sub-atomic physics. Therefore, while three bits (111) offer a single computational choice between 0 and 7, in the QC realm 3 qubits offer all eight computational possibilities at once.[126] DARPA's quantum network has been operational since 2003, offering an unprecedented level of security via Quantum Key Distribution (QKD) cryptography. Quantum-related projects for 2012 include the production of nanowires and nanotubes in support of national defense

---

2011
[124] http://www.computerworld.com/article/9221870/ Intel_s_Knights_Corner_chip_hits_supercomputing_speed
[125] The Opportunities and Challenges of Exascale Computing. Summary Report of the Advanced Scientific Computing Advisory Committee (ASCAC) Subcommittee, Fall 2010, p. 3
[126] http://www.defenseindustrydaily.com/schrodingers-contracts-us-explores-quantum-computing-03169/

applications, the former supporting QC and nanorobots programs, with the latter assisting the development of new materials and devices associated with missile defense programs.[127]

The U.S. continues to sustain the emphasis on investments in R&D, but our support on maintaining the leadership on strategic technologies is highly questionable. The corresponding compound effect is that the lack of technological leadership increases our dependence on foreign technologies, thus placing us at great risk through the supply chain threat, generated by the demands of a global economy, and aggravated by unwise acquisition policies and procedures. We should apply our efforts in regaining scientific and technological superiority in the areas that promise a greater return on investment, instead of attempting to catch up on the areas where we already conceded leadership to other nations. Asia has already captured the leadership in petascale; consequently, we should invest our efforts in achieving exascale leadership, simply because exascale is the HPC paradigm that will unveil and unlock the strategic technologies that remain unattainable at the present. The complex aerodynamic design breakthroughs required to maintain air superiority can only be achieved via the mod&sim capabilities provided by HPC exascale systems. The first global power achieving these exascale capabilities will also be the power projecting air superiority. We must remind ourselves that DARPA was created for the very purpose of avoiding strategic surprises detrimental to U.S. national security.[128]

A new aerodynamic advanced design and propulsion systems are only half of the solution, since the flying platform must overcome the exigent demands during mission maneuvers, and at the same time remaining capable of a higher degree of stealthiness to defeat the increasingly powerful and efficient adversary's radar. The reduction of RCS under mission maneuvers is also a solution waiting for the ushering of exascale HPC systems capable of delivering the computational power currently unavailable from petascale systems. The cross-pollination between the research

---

[127] http://www.smdc.army.mil/2008/TechCtr/Abstract2.pdf
[128] http://www.darpa.mil/our_work/

areas of nanotechnology and exascale computing will generate the foundation for the technological advances required to achieve technological and military superiority, with the production of new materials and advanced technology and weaponry. The one problem that remains an open challenge is the unauthorized flow of R&D data into the hands of our adversaries, because as a nation we are still incapable of protecting the confidentiality of this critical data. We invest heavily in producing this data, but we are negligent in protecting it.

The scientific, technological and strategic horizons unveiled by QC represent both a path to technological leadership, but more importantly for DoD, the path to military hegemony and strategic and tactical superiority. And while at the present time we do not have an operational QC system, we can greatly profit from the use of quantum technologies already operational in the area of quantum cryptography. The QKD schema provides a superior level of security, even transcending the dependency on current network infrastructures. Quantum teleportation over a satellite-assisted network is already under exploration, with the potential of offering not only land but sub-surface highly secure communications as well. The successful quantum teleportation of entangled photons achieved by China should serve as a rude awakening for DoD,[129] given the implications of the use of high-powered blue laser for quantum data exchange.[130] The transfer of intelligence on air, surface and sub-surface constitute a tremendous tactical advantage, and we cannot afford to take second place, since the Chinese are not the only players in the quantum teleportation race, which includes Russia[131] and Europe[132] as well.

[129] Lin Edwards, "Quantum teleportation achieved over 16 km," http://www.physorg.com/news193551675.html, May 20, 2010

[130] Matthew Luce, "China's secure communications quantum leap," http://www.atimes.com/atimes/china_business/lh26cb01.html, Aug 26, 2010

[131] First International Conference on Quantum Technologies, Moscow, http://conference.icqt.org/, July 13-17, 2011

[132] Greece WTM News, "German research brings us one step closer to quantum computing," http://www.wtmnews.gr/semiconductors-

172 THE CYBER EQUALIZER

The need for achieving leadership in HPC, both on the traditional binary path (exascale) and the quantum path (QC) is an uncontested issue, as attested by the U.S. government endorsement through the DoD HPC Modernization Program (HPCMP) and its 5 Supercomputing Resource Centers (DSRCs), including AFRL (WPAFB), ARL (Aberdeen Proving Ground, MD), ERDC (Vicksburg, MS), NAVY (Stennis Space Center, MS), and MHPCC (Maui, HI).[133] Another significant endorsement to HPC is provided by the DoE INCITE program, sustaining critical research in areas such as next-generation biofuels, nanotechnology, astrophysics, nuclear fusion energy, and aeronautical engineering, among others.[134] The only issue is the type of microprocessors architecture to employ in order to overcome the challenges of achieving petascale performance within viable parameters for power consumption and heat dissipation, and developing faster optical interconnects and enhanced algorithms for optimized utilization of available computing processing cycles. However, the ultimate HPC is the achievement of quantum computational power, under conditions transcending the limitations of current microprocessors. Significant and promising achievements toward this goal have been recently reported by researchers at the Max Planck Institute of Quantum Optics[135] and at the UC Santa Barbara.[136] We hope this is the prelude to capturing the leadership in HPC.

07/6158-German-research-brings-us-one-step-closer-to-quantum-computing.html, 22 March 2011

[133] DoD High Performance Computing Modernization Program, http://www.hpcmo.hpc.mil/cms2/index.php, 20 June 2011

[134] U.S. Department of Energy, "2012 INCITE Call for Proposals," https://hpc.science.doe.gov/allocations/calls/incite2012.

[135] Max Planck Society, "Max Planck Society (MPG) Research News," http://www.research-in-germany.de/67310/2011-05-03-single-atom-stores-quantum-information,sourcePageId=12482.html, 5/3/11.

[136] ScienceDaily, "Physicists Demonstrate Quantum Integrated Circuit That Implements Quantum Von Neumann Architecture," http://www.sciencedaily.com/releases/2011/09/110901155259.htm, Sep. 2, 2011

Addendum. June 2012 TOP500 List

Two weeks after submitting this book for registration to the United States Copyright Office, and important development took place in the TOP500 HPC list.[137] On 18 June 2012 this author delivered a presentation on "Exascale and Quantum Computing" at the Cyber Defence 2012 Conference in London, and upon completing this presentation, several members of the audience approached this author to share the news received on their laptops, tablets and smart phones regarding the new alignment on the top five HPC systems. The news announced the United States had recaptured the leadership on HPC, and the two previous leaders, namely Japan and China, had taken the second and fifth place in the new 39th official TOP500 List, released on 18 June 2012 at the International Supercomputing Conference in Hamburg, Germany.

The top five HPC systems in the world are now arranged as follows:

Rank	HPC name	Performance	Country
1	Sequoia	16.32 Pflop/s	US
2	K	10.51 Pflop/s	Japan
3	Mira	8.15 Pflop/s	US
4	SuperMUC	2.89 Pflop/s	Germany
5	Tianhe-1A	2.57 Pflop/s	China

By comparison, the previous November 2011 top five list was dominated by the Asian presence with two HPC systems from Japan, two from China, and only one system from USA, ranking

---

137 http://www.top500.org

in the third place. The new distribution of the current top five HPC systems represents a dramatic change on HPC leadership, bringing a more balanced distribution of HPC prowess between the West and the East. This is very important because HPC capabilities translate into leadership and superiority in the various critical fields of science, medicine, engineering, and defense industries.

INDEX

www.ingramcontent.com/pod-product-compliance
Lightning Source LLC
LaVergne TN
LVHW042335060326
832902LV00006B/183